SPACEWALKER

"As an astronaut for over three decades, one who participated in some of the most challenging and exciting Shuttle missions, Jerry Ross knows spaceflight. In *Spacewalker*, he details the exultation of actually being in space and brings to life the realities of preparing for and executing one of the most difficult of all human endeavors. *Spacewalker* is the book for anyone who ever dreamed of flying in space."

–Neil Armstrong, Commander of Apollo 11
(June 2012)

"I have known Jerry Ross for many years, as an astronaut, a friend, and as a fellow alumnus of Purdue University. We've had many opportunities to share our experiences and our dreams for space exploration. We share a passion for inspiring young people. This is the story of one man's lifelong quest to explore the unknown, overcome setbacks and obstacles, and keep the beacon of space shining in the hearts of all people, young and old. This book is about an American Dream."

–Eugene Cernan, NASA Gemini and Apollo astronaut;
co-author of *The Last Man on the Moon*

"Seven times in space, a first and a record matched by only one other. As Flight Director I sweated out the gut-wrenching final seconds on launch day many times. In the seconds before launch I said a prayer for my team, and then at the instant the rocket was freed of its earthly shackles I briefly prayed for our crew. Most astronauts of my day would fly but two or three missions, but Jerry Ross faced the fire and the risks of launch day repeatedly. This book is the story of a common man from the Midwest who became an American hero . . . a model for the youth of our nation and for those who will accept the challenge to follow in his footsteps."

–Gene Kranz, NASA Flight Director for Gemini and Apollo,
including Apollo 11 and 13; author of *Failure Is Not an Option*

"As a NASA training manager, I knew Jerry Ross as one of the steadiest and smartest astronauts in the agency. He was also one of the nicest! No matter who you were or what your job was, Jerry always took the time to express his appreciation for the work you were doing. Frankly, considering the pressures we were always under, I wondered what made Jerry so nice. Now that I've read his memoir, I know! Jerry Ross is the genuine article, a man of humble roots who decided to be the best he could be while never forgetting the lessons of honesty, integrity, and family he learned along the way. Great memoirs have to strike a balance between history, circumstance, purpose, and passion. By this or any standard, *Spacewalker* is a great memoir. Jerry tells his story straight from the heart, and it is a magnificent tale of courage and faith. If you liked *Rocket Boys/October Sky*, you're going to love this book!"
    –Homer Hickam, author of *Rocket Boys/October Sky* and *Crater*

—————————————————— ● ——————————————————

"This book is a must-read! Astronaut Jerry Ross has seen our Space Shuttle program from start to finish, and more. No other person has experienced the depth and breadth of the Space Shuttle intrigue as Jerry has. You will be inspired by the story of a young man chasing his dream . . . and achieving it many times over. You will be taken inside the private side of astronauts' lives, as they train and fly their intense and risky missions. You will learn and remember the greatness of America's Space Shuttle program. This story has all of it: the humor, the tragedy, the family view, the technical side, as well as Jerry's personal perspective of our country's great Space Shuttle program. There is no other book like this; it is a gem!"
                    –Eileen Collins, first woman to pilot and command
                                              US space missions

SPACEWALKER

MY JOURNEY IN SPACE AND FAITH AS
NASA'S RECORD-SETTING FREQUENT FLYER

BY **JERRY L. ROSS**
WITH **JOHN NORBERG**
WITH A FOREWORD BY **EUGENE CERNAN**

Purdue University Press ● West Lafayette, Indiana

Ross, Jerry Lynn, 1948-
  Spacewalker : my journey in space and faith as NASA's record-setting frequent flyer / Jerry L. Ross with John Norberg.
       p. cm.
  Includes index.
   ISBN 978-1-55753-631-0 (pbk. : alk. paper) -- ISBN 978-1-61249-232-2 (epdf) -- ISBN 978-1-61249-233-9 (epub) 1. Ross, Jerry Lynn, 1948- 2. Astronauts--United States--Biography. 3. Outer space--Exploration--United States. I. Norberg, John. II. Title. III. Title: My journey in space and faith as NASA's record-setting frequent flyer.
  TL789.85.R67R67 2013
  629.450092--dc23
  [B]
                                    2012029105

—————————————————●—————————————————

With all of my love to Karen, our children Amy and Scott, our daughter-in-law Faith, and our beautiful granddaughters Cassidy, Katie, and Emily. May we all walk each day in God's grace.

—————————————————●—————————————————

CONTENTS

FOREWORD

In the spring of 1961, I was a young US Navy officer when President John F. Kennedy challenged the people of the United States to look to the Moon. It changed my life.

But more than my life, it changed the nation. It changed all of us. It wasn't just a call to land a man on the Moon and return him safely to Earth. It was a challenge to show the world the strength, commitment, and capabilities of a united American people. It was a challenge to believe in ourselves and discover the unlimited potential within us to accomplish seemingly impossible goals. It was a challenge to our national spirit, a call to look upward and to dream.

We succeeded. In December of 1972, I stood on the dusty surface of the Moon, looked back at the blue ball that is our Earth, and became part of the realization of a dream that had been within people for all time.

I was the last Apollo astronaut to walk on the Moon, but America's space program didn't stop with Apollo. The Moon was never the end goal. The Moon was always the beginning. Reaching the Moon was just the first step in opening the universe to human exploration and understanding.

We have since flown a space vehicle capable of returning to Earth and going back into space again. From space, we have launched satellites, which bring communication and education to every corner of the world, helping to unite people everywhere. We have placed observatories in space that are seeing to the beginning of time, bringing humans new understanding as well as new mysteries to solve. We have worked in peace with other nations to build and operate an International Space Station. And we are still just at the beginning of this incredible journey.

NASA has led our nation and the world in the most exciting and productive fifty-five years of exploration in human history. Everywhere on this planet people know what NASA is and what it represents.

NASA is more than a space agency. It stands for the greatness that is America and the potential we have for even more. It represents to the world the best of human excellence, integrity, and innovation.

At times NASA has failed, and the results have been tragic. At those moments it would have been easy to quit, but we came together as a nation, learned from our mistakes, and sent NASA forward again to explore new and exciting possibilities.

Now we are at a crossroad. With the retirement of the Space Shuttle in the summer of 2011, NASA lacks clear goals. We are walking away from all that has been accomplished and leaving it for others to take the next giant leaps forward.

If America abdicates leadership in space, not only is human spaceflight and space exploration at risk, but I believe the future of this country and the future of our children and grandchildren are compromised as well. If we stop reaching into the unknown and in the process lose generations of acquired knowledge in space science, engineering, and technology, how will we regain what we've lost? Once abandoned, what will become of the American spirit to explore?

Accomplishments in space have inspired generations of young people to go farther in their lives than they ever dreamed possible. I talk to students often. They tell me that, even as we seem to have lost our direction in our goals for space exploration, they still want to visit Mars. They want to explore space, and they want to expand the boundaries of human knowledge. They want to go where no one else has gone and do things no one else has ever done before.

And they will. If we give them the tools—education, inspiration, and opportunity—if we pass the dream to them, this generation of Americans will take us farther than we have ever been, in space and here on Earth.

This is not just about space. It's about us. It's about our nation. It's about what President Ronald Reagan called the shining city on a hill that is a beacon of hope for all the people of the world.

I have known Jerry Ross for many years, as an astronaut, a friend, and as a fellow alumnus of Purdue University. We've had many opportunities to share our experiences and our dreams for space exploration. We share a passion for inspiring young people.

Spacewalker: My Journey in Space and Faith as NASA's Record-Setting Frequent Flyer is the story of one man's lifelong quest to explore the unknown, overcome setbacks and obstacles, and keep the beacon of space shining in the hearts of all people, young and old.

This book is about an American Dream.

Eugene A. Cernan
Captain, United States Navy (Retired)
Pilot Gemini 9
Lunar Module Pilot Apollo 10
Commander Apollo 17

PROLOGUE

We grew up together in the 1950s surrounded by fields of wheat, corn, soybeans, and hay in northern Indiana.

The mills of Gary were just fifteen miles north of our homes, gushing black smoke out of tall blast furnaces, making steel to build an America that was bursting at its seams after World War II.

But all that seemed far away from the rural countryside where my best friend, Jerry Ross, and I played catch, hit baseballs, and explored woods and marshes. We lived a quarter mile apart on a quiet county road, and there was nothing in between but rich Indiana farmland. In the summer, the sky above was big and blue with hawks gently gliding, circling, searching for prey. Red-winged blackbirds, robins, and turtle doves filled the air with their calls.

We lived far enough from the city lights that the sky became very dark at sunset and filled with stars—bright, twinkling guideposts that dotted the blackness. You don't see starlit skies like that very often anymore, unless you are in the mountains or sparsely populated areas. But in the 1950s, in the nighttime sky over Indiana, you could see enough stars to assure any boy there was something out there that was way beyond our understanding.

On summer evenings, with nothing more important to do, Jerry and I would gaze at that starry sky. We lay on stacks of freshly baled hay in his grandfather's field beside Jerry's house. The hay bales were hard, but they were comfortable. They were more fun to lie on in the evening than they had been to stack earlier in the day, when we worked under the hot sun, riding on a wagon behind the baler through the muggy, buggy fields of Lake County with dust and chaff flying into our faces.

Other than the smell of freshly baked homemade bread, there isn't a better smell in the world to me than new-mown hay. In old songs about Indiana it's called "the breath of new-mown hay" that ". . . sends all its fragrance through the fields I used to roam." I know that feeling. It was part of my childhood.

Even today that sweet, aromatic, warm scent makes me feel comfortable and secure. Jerry and I grew up together, comfortable and secure. Memories from that time—the feel of a soft breeze at sunset, the starry nighttime sky, and the full aroma of freshly cut hay—still stir my emotions and make me feel like a boy again.

Lying on those bales looking up at the sky, Jerry and I never spoke much. There wasn't much that needed to be said in the awesomeness of what we saw. The stars shimmering above made a statement so powerful that talking seemed trivial.

But one night Jerry did speak. It's strange that I can remember it after all these years, but I can. Maybe somehow even then I suspected it wasn't just a moment when a boy was expressing an impossible dream. It was a moment when a boy was looking at his future, off in the distance, deep in the sky.

"You know what, Jim? Someday people are going to go up there," Jerry said.

I wasn't so sure.

"Yeah, right," I replied.

"I'm serious," he said. "One day people are going to go up there. And I'm going to be one of them."

In the 1950s many boys growing up in rural, northern Indiana wanted to be farmers. Some wanted to work in the steel mills where there was good money to be made. Others had big dreams to play professional baseball. Some had ambitions of becoming lawyers, doctors, or even cowboys out west.

But I only knew one boy from those early days of my life who wanted to go into space. And as far back as I can remember, that was his only dream.

Jim Gentleman
Crown Point, Indiana

JERRY L. ROSS

- First person in history to be launched into space seven times.
- With NASA Astronaut Franklin Chang-Diaz, holds the record for the number of times launched into space—seven.
- One of only three NASA astronauts to support the US Space Shuttle program from before the first launch through its final landing.
- Supported the International Space Station program from inception through final assembly and initial operations.
- Led the team of spacewalkers who began assembly of the Space Station and among the first to enter it on orbit.

- With nine spacewalks, is tied for third among space-walkers worldwide and second in the US for most spacewalks performed.
- With over fifty-eight hours on spacewalks, is third among spacewalkers worldwide and second in the US for time spent walking in space.
- Helped develop and create the facilities, tools, techniques, and training needed to construct the International Space Station.
- Performed a leadership role in the recovery of the Space Shuttle *Columbia* wreckage.
- Awarded two Defense Superior Service Medals, the Air Force Legion of Merit, four Defense Meritorious Service Medals, two Air Force Meritorious Service Medals, the National Intelligence Medal of Achievement, and fifteen NASA medals.
- The only three-time recipient of the American Astronautical Society's Victor A. Prather Award for spacewalking achievements and the only four-time recipient of the Society's Flight Achievement Award.
- Named to the Purdue University Air Force ROTC Hall of Fame.
- Recipient of the honorary doctor of science degree from Purdue University in 2000 and the Distinguished Engineering Alumni Award from the Purdue University College of Engineering in 2004.
- In 2003, his hometown school system, Crown Point Community School Corporation, named Jerry Ross Elementary School in his honor.
- A member of the Silver Team, NASA's first pair of spacewalking grandfathers.

ONE

Sputnik, *a mouse, and blackberry pie*

When I was little, some kids said they wanted to be farmers, lawyers, teachers, doctors, baseball players, or firemen. But I couldn't say I wanted to be an astronaut.

When I was born on January 20, 1948, there were no astronauts. There were no space capsules or Space Shuttles. There was no NASA and no International Space Station. Orville Wright—who with his brother Wilbur designed, built, and flew the world's first airplane—was still alive. He died ten days later at the age of seventy-six.

Airplane flight was barely forty-four years old, and the term "Jet Age" was so new it was rarely used. The real "Jet Age" was still a few years away, and the "Space Age" would drop right on top of it as daring pilots such as Chuck Yeager pushed the boundaries of possibility. On October 14, 1947, Chuck had piloted the Bell X-1 airplane and flew faster than the speed of sound for the first time.

The only spacemen in 1948 were Buck Rogers and Flash Gordon, and they were in comic books and movies. While space had captured the imaginations of science fiction writers and some scientists, the public viewed spacemen as whimsically as the "Man in the Moon." Space travel was still a fantasy.

In 1948, Virgil "Gus" Grissom, who later would become the second American in space and the first person in the world to be launched twice, was a sophomore studying engineering at Purdue University, seventy miles south of my northern Indiana hometown. He'd probably never heard the word "astronaut." He wanted to be a United States Air Force pilot. Neil Armstrong was an eighteen-year-old Purdue freshman with no idea that in twenty-one years he would put his bootprints on the Moon.

While they could not know what lay ahead, the future was brewing inside them as they walked across the redbrick campus carrying thick calculus and physics textbooks and slide rules holstered in leather cases. The next twenty-one years would transform the world, and I lived through it. It made me what I am today.

I am an astronaut.

In my life I have seen views of Earth and the universe from a perspective once known only to God. And yet, I grew up in a time so distant from today that electric can openers were considered high tech. And my family didn't have one.

I have launched into space seven times and ventured into the blackness of the universe on nine spacewalks.

From two hundred miles high I have watched lightning pop through dark clouds stretched across the Amazon, seen the Himalayas reach up to greet me, and looked down at the Indiana hometown from where I once looked up at the stars.

The world we live in today was unimaginable when I was a boy, and space has made the difference. In the twenty-first century, all of us are as closely linked to space as the cellular phones in our pockets. We are linked to space by television, financial transactions, the Internet, weather forecasts, and GPS. The technology we employ every day has been made possible by space systems. Our exploration of space has not only opened up the universe and fostered new technologies that we use here on Earth, but it has also taught us about our planet and its environment. The impact has been revolutionary.

Thirty years of that exploration was dominated by a ship that was launched, returned to Earth, and sent back into orbit again—the Space Shuttle. We were awed by it, inspired by it, and sometimes saddened by it. And I was there for it all—from the first Shuttle mission to the last. It was the adventure of my life. Now the Space Shuttle program has run its course, the orbiters are in museums, and I am very concerned that America is walking away from its hard-earned leadership in space.

Our missions were filled with danger, excitement, hard work, tension-breaking laughter, and good times as we launched off the face of the Earth in an explosion that could be seen, heard, and felt miles away. We did things no one had ever done before. I want to share the amazing experiences I had pursuing my dream and help you understand what it felt like to ride those rockets and walk in space.

We saw views that no camera has ever been able to fully capture. Seeing the beauty of Earth from orbit along with the enormity, the complexity, and the order of the universe strengthened my faith in a loving God who gave us a beautiful, fragile place to live. I want to share how God has directed my path with the hope that others will choose to experience their own journeys in faith.

My journey wasn't easy. It was difficult. It came with all the failures and setbacks that are part of life. I reached my dreams by always believing in them—no matter how distant they seemed—and by sticking with them through trial, error, and success. I want you to know that the incredible dreams inside you, whatever they might be and wherever they might take you, can be within your grasp if you believe and persevere.

My journeys in space, faith, and life all began at the same time, in the same place, in the all-American Midwest town of Crown Point, Indiana, in an era when the Moon was still unreachable but a boy's dreams could take him to the stars.

I was born at Methodist Hospital in Gary, Indiana, fifteen miles north of Crown Point. I was six weeks premature, skinny, and blue. The nurses called me "Squeaky" because of the noises I made. I had pneumonia twice that first winter, and some of my relatives didn't think I would survive.

My father's name was Donald John Ross and my mother was Phyllis Elaine (Dillabaugh) Ross. They were married on September 28, 1946, in Leroy, Indiana, a very small community a couple of miles from their homes, where there was a grain elevator along the railroad. Growing up, Mom and Dad lived in the same neighborhood and attended the same church, Leroy Methodist, where they were married.

I don't know what dreams my parents had for me when they brought me home from the hospital to a rental house outside of Crown Point. They never said. They just gave me love, a firm Christian foundation, and a strong sense of family and community.

I was born with God-given talents that ultimately helped me fly into space. I had common sense, a mechanical bent, a strong body, and natural aptitudes for science and math—as long as I studied hard. And I was born with a curiosity about how things work, about what was in the woods and wet marsh a mile from my house, about what was out there in the sky beyond our knowledge and imagination. I wanted to know.

My dad was a central figure in my life. Dad had a great laugh and exercised it frequently. And he was a very strong guy. He was one of the few people I ever knew who could do one-arm push-ups. His athletic prowess was recognized by his high school football and basketball teammates when they gave him the nickname "Barney" after the famous 1930s champion boxer Barney Ross.

Dad was a very good athlete in high school and had been offered a football scholarship to Michigan State, but he didn't take it. I know he would have loved playing Big Ten football. It might have been his dream come true. But that was a time when people didn't generally go to college, not farm kids from Crown Point anyway. In those days, in the rural community where he grew up, you graduated from high school and went to work. Instead of going to college, he went straight to the steel mills. I often wondered how a different decision might have changed his future and mine, or if I would have ever been born!

Dad was drafted into the US Army during World War II. He served in the South Pacific, and for four years he never once got to come home. He and Mom had dated before he was inducted, and they wrote letters back and forth throughout the war. Dad never said much about his war experiences, but we did know he was away from home for a long time. Growing up, knowing things like that about my father and mother helped me put any hardships I had into perspective.

Dad's parents were Joseph and Mabel Ross. We kids called them Grandpa and Grandma Joe. Grandpa Joe had given my folks three acres of land off the corner of his farm, and that's where Dad, with some help from family and friends, built the house where we lived. The house was wood frame, one story, and it was covered with white asbestos shingle siding. It was less than eight hundred square feet, with just two bedrooms, one bathroom, a living room, a small dining room, and a kitchen. That was it. There was a basement underneath the house with a coal furnace, a water heater, and a washing machine. There also were clotheslines and a fruit cellar down there.

It doesn't sound like much now, but I never knew it was a simple, small house. I thought it was the greatest home a kid could ever have. I still do. The house didn't have a lot of space, but it was full of love.

I was two years old when my sister Judi was born and five when my parents had my sister Janet. When Dad replaced the coal furnace in the basement with one that burned fuel oil, he converted the coal bin into a bedroom for me. I loved it. We didn't have air conditioning, so in the summer the basement was the coolest place in the house.

Our house was on the Old Gentleman Road within a mile or so of both sets of grandparents. From our windows we could look southeast across the fields and see Grandma and Grandpa Joe's house and farm buildings. Grandpa Joe's ancestors had come from northern Ireland in 1849. Grandma Joe's mother was a Gibbs, and the Gibbs family entered North America in about 1630 from England. Through my grandmother's family I'm related to Orville and Wilbur Wright, and maybe that's why I grew up looking to the sky.

My mother's parents, Ellis and Lora Dillabaugh, lived over on the highway north of us. Grandpa Dillabaugh's German-speaking Swiss ancestors arrived in North America in 1710 and settled in the area that became New York State. Grandma Dillabaugh's Schaaij and Arentze families had come from the Netherlands to the open plains south of Chicago in 1847.

Dad worked nearly all his life at United States Steel in Gary, sometimes six or seven days a week, ten to twelve hours a day. He worked in the blast furnace maintenance division, and for a long time he worked as a rigger. The riggers used hooks, cables, block and tackle, and muscle to lift and move heavy equipment and to perform maintenance to keep the molten iron flowing. Later Dad became a foreman in charge of the group that kept everything oiled and lubricated to prevent mechanical breakdowns.

It was hot, dirty, and smelly work that could be dangerous. He was around high-power, high-voltage electric systems, hot metal, and poisonous gasses. Normally he would take a shower at the steel mill before he came home, and I can remember the filthy work clothes he gave Mom to wash. When we got a new washing machine, Mom kept the old one and used it exclusively for Dad's work clothes, covered with coke dust, iron ore dust, and grease.

As most mothers did back then, Mom stayed home with us until I was in sixth grade, and then she worked for several years as a secretary for the Center Township Trustee. She later became a secretary in our school system. Mom ran the house, but we also knew if we didn't straighten up when she told us to, we were in for it when Dad came home. He was the enforcer. My folks made us behave but they allowed us to be kids, and I have very good memories of my childhood, being a tow-headed boy running around in the huge yard.

Mom nurtured and encouraged my sisters and me and helped us to grow up not being afraid to try things we might otherwise have considered beyond our reach. One of my earlier memories is Mom's encouragement when I become infatuated with space. By the mid-1950s people were beginning to talk about launching a satellite and about human beings traveling into space, which really captured my imagination.

My aunts and uncles knew it, and they would save their *Look* and *Life* magazines or anything with an article about space in it for me. I'd cut out the articles, read them, and study them. Mom helped me paste the articles into scrapbooks. She typed captions, which we put in the books as well.

I wish I had those scrapbooks. When I left for college, they were stored on the shelves in the fruit cellar. Unfortunately, all those scrapbooks were ruined by the dampness in the basement, so Mom eventually had to throw them away, along with my baseball cards.

Mom was always very fair. Many years later, after I'd launched on the Space Shuttle, people would ask my mother if she was proud of her son, the astronaut. Her answer was always immediate and always the same: "I'm proud of all my children." She meant it, and we knew it.

Crown Point was a great place to grow up. It was a town where you learned the value of hard work and the importance of faith. Families were strong, and morals and ethics were part of the fabric of the community. They were emphasized in school, as important as math and science. We learned that God came first, then your family, and then your country. And the first need of your family was having you there—to play ball, to go to church, to eat dinner, to pray with and kiss goodnight. As much as science and math, all of this was the foundation that helped me become an astronaut.

A child could roam the neighborhood and surrounding farmland and not be afraid that anything bad would happen. In the summer we played outside, and our mothers did not expect us back until lunch and after that not again until dinner. We were free to explore. The whole community watched out for us. If we misbehaved, the word got back to our parents faster than we could get home to defend ourselves.

When I was only about four years old, I disappeared. Mom called my grandparents and said, "Jerry's gone. Have you seen him?" Grandma Joe looked out the window and replied, "Here he comes. I can see his head above the corn." It must have been pretty early in the summer for my young head to stick up above the corn. I can still remember emphatically being told that it wasn't a good idea for me to go to Grandpa Joe's by myself, so I didn't do it again—until I was five.

My best friend, and the only other kid within shouting distance of my house, was a boy more than a year younger than me named Jim Gentleman. He lived a quarter mile down the road on the opposite side.

We played many, many days in the countryside around our homes. He's still my best friend. When Jim and I first met, we were young enough that we were only allowed to play in his yard or mine. Most of the time when we played in my yard, we could be found in my sandpile. We used our farm toys to plant and harvest imaginary crops, or we used our "heavy equipment" to construct roads. We also spent hours playing with my four-legged best friend named Prancer, a black cocker spaniel who was my constant companion until his death in 1962.

------------------------------------●------------------------------------

JIM GENTLEMAN: *When I first met Jerry, he had a little pedal tractor. It was a red Farmall. He would come down the road to my house lickety-split with his father chasing behind him. His father used to say, "This kid is getting harder and harder to keep up with."*

Jerry's dad, Don, worked very hard, and his mother was the kindest person you'd ever meet. She couldn't do enough for people. It was great to have a friend who had nice parents, and our parents were friends, too. My sister Cathie was a close friend of Jerry's sister Janet.

I became an osteopathic physician, and I was Don's doctor until the day he died. He never gave up. That's something I remember about him. Don wouldn't be feeling well, and he'd tell me, "Getting old isn't for sissies," but he'd keep going. That's where Jerry got his determination.

There was a culvert that ran under the road about halfway between Jerry's house and mine. Jerry was always looking to explore, and he was always looking for adventure. The first time he saw that culvert pipe he said, "Come on, Jim, follow me." And he crawled in. We could hardly fit through it, but I followed him. I always followed Jerry—sometimes right into trouble.

------------------------------------●------------------------------------

Jim's dad, Big Jim, worked in the Chicago stockyards. Once he brought my dad a young calf that had been born while its mother was being shipped to the yards. The people at the stockyards didn't have a good way to deal with the calf, and they wanted to find it a home. Dad thought the calf would give him a good chance to teach his children responsibility and how to take care of an animal. He moved a small building from my Grandpa Joe's into the field right behind our backyard and built a fence to give the calf a grazing area. The calf was very wild and routinely jumped the fence or just ran through it, so Dad often had to chase it all over the neighborhood.

One time Dad caught the calf in our backyard and wrestled it to the ground. Exhausted, he got on top of it and started pounding his chest like Tarzan. At the same moment, the calf gave a great spurt of effort, sprang to its feet, and threw Dad head over heels onto the ground. So much for the calf, and so much for us kids learning responsibility by feeding it. Dad took the calf back to Big Jim, where it joined the Gentleman herd. Maybe it was Dad who learned a lesson that time.

When I was a kid, we had a party-line telephone system, sharing a phone line with neighbors. I still remember our number—8053-L1. Grandma Joe loved the party line because she could listen in on other people and catch up on everything the neighbors were doing. I remember times when Dad and I walked into my grandparents' house and Grandma Joe had the telephone to her ear, quietly listening to what people were saying on the party line. Dad would get her goat by loudly saying, "Mom, how are you doing?"

She'd whisper, "Be quiet!" with a red face, as she carefully hung up the phone.

In addition to our grandparents, we lived near aunts and uncles, and I grew up surrounded by cousins. Christmases were filled with family and love. On Christmas Eve we would go to Grandma and Grandpa Dillabaugh's home, and almost everyone in the family was there. It was a madhouse. Mom had a half-brother named Grant who had four boys and another brother named Don with four kids. Brother Ellison had three, and the last brother named Lynn (Uncle Itch to us) had six kids. That added up to a lot of noise and commotion on Christmas Eve. All of us kids loved black olives, and Grandpa Dillabaugh bought large cans of them. We put the olives on the tips of our fingers and walked around the house eating them.

Grandma Dillabaugh was the greatest baker who ever lived. She made some of the most delicious pastries at Christmas. I can still taste her molasses cookies with white sugar frosting. She baked long johns and bear paws, and she made an apple slice sheet pastry that melted in your mouth.

On Christmas Day we'd go to Grandpa and Grandma Joe's house. Dad's sister Margaret had four daughters, his brother Harold had four sons, and his twin sister Dorothy had two boys and two girls.

Dad's other sister, Aunt Marian, loved to sing, and she liked things very structured. Every Christmas she planned a program in which we all had to sing. All of us kids rolled our eyes then, but these are good memories to laugh about now. On Christmas afternoon the adults played pinochle. I'd sit and watch them, and when I got old enough, the adults invited me to join them. That was a big deal!

My family always went to Sunday school and church services at the Methodist Church in Leroy. Two of my great-great grandfathers, Samuel Love, Sr. and Harvey Gibbs, a Civil War veteran, were founders of the church when it was built in 1888. Sam Love donated the lot where the church still stands. Two additions were made to the original building, and family members—including my parents, me, and my Grandpa Dillabaugh's house moving company—helped with those projects. I still like the good old church hymns we sang at Leroy Methodist: "I Love to Tell the Story," "Christ the Lord is Risen Today," "This is My Father's World," "The Old Rugged Cross," "How Great Thou Art," and "Are Ye Able, Said the Master."

Crown Point was just fifty miles from downtown Chicago. In the summers our family would go to the Field Museum of Natural History and the Museum of Science and Industry. And we went to the Brookfield and Lincoln Park Zoos.

Dad would also take me to Gary to a great ice cream parlor attached to a dairy, where they had a delicious menu item called Tub-O-Malt. The Tub-O-Malt was at least a half-gallon of ice cream made into a malt. Dad and I each had our own. We didn't split. It was a real guy thing. A Tub-O-Malt was a challenge, but I welcomed it. I was a little guy trying to chug that all down. It was great!

But my favorite thing to do with my dad was to go to Wrigley Field. He was a die-hard Cubs fan. The dictionary definition of "die-hard" is a person who stubbornly clings to and tenaciously defends a seemingly hopeless or lost cause. That was Dad, and therefore, me!

Chicago is a city divided when it comes to baseball, and the dividing line runs east/west. For the most part, Cubs fans live north and White Sox fans live south of the line. Since we lived south of Chicago, most people in Crown Point were Sox fans. Dad's brother, Uncle Harold, lived on the opposite corner of Grandpa Joe's farm, and he was a big Sox fan. All four of his sons—Keith, Lanny, Dean, and Kent—are Sox fans. And my best friend, Jim, is a Sox fan. Dad and I were surrounded, but we were die-hard Cubs fans.

Dad and I went to Wrigley Field two or three times a summer. Those were always anticipated and wonderful days making lifelong memories with my dad. We never ordered tickets. We could buy them on game day. I got to see Ernie Banks and stars from other teams— Hank Aaron, Eddie Mathews, and Don Drysdale. We would sometimes take Jim along so he could see a good baseball team play!

I especially remember one weekend with Dad at Wrigley Field. There was a doubleheader on Saturday, and Dad and I had so much fun that we decided to go back for a second doubleheader scheduled for Sunday. Actually, we were both relieved when the second game on Sunday was rained out because our rear ends were getting sore about the time the first game was over.

After my first Space Shuttle mission, STS-61B in 1985, I was invited to Wrigley Field to return a 1984 Divisional Championship pennant that I had taken with me into space. After all the years of Dad taking me to the ballpark, that day I took him to Wrigley. Before the game I went out onto the field and presented the banner to Jody Davis, the Cubs catcher. Dad and I got to go into the dugout and the locker room. We met several Cubs players and the Cubs General Manager Dallas Green. Dad was in heaven. I think it was one of the greatest days of his life. Seeing the smile on his face made it one of my greatest days, too!

————————————●————————————

JIM: *Jerry and I both loved baseball, and his father always hit fly balls to us in the yard. He didn't have to hit the balls very far to make it seem like a long fly to kids like us who were eight or nine years old. Jerry's dad always seemed like Superman to me.*

Later he was coach of our Babe Ruth team. I was a pitcher and only thirteen years old. Jerry was the catcher, and he was already fifteen, in his final year. I was just starting, and it was a big jump from Little League to Babe Ruth. Jerry's father continued to play me even though in batting I went 1 for 51. The only hit I got was off the best pitcher in the league. I could never figure out why Jerry's dad played a guy who was 1 for 51. Years later I asked him about it. He said he always thought I would be a good player. He just wasn't sure exactly when.

I did get better. I played Big Ten baseball for Purdue until I had to give it up for premed studies.

————————————●————————————

Baseball was an important part of my life growing up. I had a lot of good times playing ball, except one. During the summer between fourth and fifth grade, we were in our front yard and Dad was hitting fly balls to Jim and me. Jim hadn't brought a baseball glove with him, so I gave him an old catcher's mitt to use. A catcher's mitt is not the best thing to use to catch fly balls, and that glove wasn't very good. Dad hit a high fly and Jim got into position to catch it. I ran behind him to back him up, just like the pros do. When Jim reached up, the ball hit the top of the catcher's mitt. The ball deflected off his glove at an odd angle, sailed over the top of my glove, and hit me right in the mouth.

I was crying and bleeding, and my mom came running out of the house. Moms always know when a cry really means trouble. I had a swollen lip, blood pouring from my mouth, and two broken teeth. We couldn't find the missing pieces of teeth and assumed I had swallowed them. Mom took me to our dentist, Dr. Tullis, and he put temporary caps on my teeth.

The two broken teeth had cut a gash in my lower lip, and after a week or so the lip was not healing. During a second visit to the dentist, an X-ray revealed that the two missing parts of the teeth were still in my lower lip. Mom and I left the dentist and went straight to see Dr. John Birdzell. Dr. Birdzell was a great guy and had been our family doctor for years. He had delivered me.

We explained the problem to Dr. Birdzell, and he took us back into his treatment room. He injected some Novocain into my lip and then started talking with my mom, catching up on what was new without noting the time of the injection. He returned to me with a pair of scissors before my lip felt numb. As he cut the inside of my lip open, I could feel Novocain gush out and rush down my throat. Dr. Birdzell proceeded to fish out the two pieces of teeth. It hurt so much, but I couldn't say anything because Dr. Birdzell had his hands in my mouth.

We were out of the office and on the way home before my lip and throat became numb. I told my mother what had happened and got the sympathy I deserved. "Next time," she said, "don't catch the baseball with your mouth."

When I started school in first grade I went to Washington School, the same country school my folks had attended. There were about thirty kids in a grade. My first grade teacher, Mrs. Sigrid Isolampi, died partway through the year, and Mrs. Jeanne Steeb, who had taught my parents, came out of retirement and took over. During the next twelve years, I had several teachers who had been teachers for my parents.

Our family lived in the southeast corner of the township, and normally my sisters, Judi and Janet, and I were among the first ones picked up by the bus in the morning and the last ones dropped off at night. It was a long day for kids, but we had fun on the bus. Once in a while I'd get in trouble with the driver—just doing things it was strongly suggested that I not do. I teased girls—a lot—which continues to be a favorite pastime of mine. Just ask my granddaughters!

I know I gave my sisters a particularly hard time. I called Janet "Jan Nut," and I loved to beat Judi and Jan Nut in Monopoly and rub it in. In those days I believed it was the sacred duty of a brother to make life difficult for his sisters. Now Jan Nut gets even by calling me "Astro Nut."

Our family watched Walt Disney's *Disneyland* television program on Sunday nights. Back then, families only had one TV in a house and everyone watched together. One Sunday in 1955, Walt Disney had rocket scientist Werner von Braun on *Disneyland* to talk about launching rockets into space. The show was incredible. It included scientists and cartoonists teaming to explain rocket propulsion and space travel in a way that both children and adults could understand and enjoy. Von Braun used the show to try to excite the public about his rockets. He sure excited me.

The Space Age officially arrived in 1957.

The Cold War between West and East was in full swing by that time, and the United States and the USSR were stockpiling nuclear weapons. In August of 1957, the Soviet Union launched the first intercontinental ballistic missile (ICBM), which demonstrated that they could deliver nuclear bombs enormous distances. There was a great deal of fear about nuclear war.

I entered the fourth grade in the fall of 1957, and my teacher was Mrs. Effie Laney. Mrs. Laney was special. She encouraged all of us, and she helped light a fire in me. She encouraged me to grow and to do whatever I wanted to do. She was the kind of teacher you wanted to make happy. She made you want to learn, and she made learning fun.

The Russians launched *Sputnik 1* on October 4, 1957. It was a day that changed the world. *Sputnik* was the first artificial satellite to be put in space, and it circled the globe about every ninety minutes, just like our US Space Shuttles would twenty-four years later. *Sputnik* launched the Space Race, and the Space Race became a major part of the Cold War.

When the Russians beat us into space with *Sputnik*, it frightened people in America. Folks in Crown Point and all around the country read newspaper stories about the beeps being transmitted from *Sputnik* and what they might mean. Were they secret signals going back to Moscow containing vital information about the United States that had been gleaned from space? Actually, the beeps came from a one-watt radio and didn't mean anything, but we didn't know that in 1957.

Many people spent time in their yards looking up at the night sky trying to spot the satellite as it passed overhead. What most people sighted wasn't the tiny satellite. The flashing light they saw was actually the tumbling, spent second stage rocket booster that was also in orbit. I never saw either one. But I looked. Even if it was the Soviets who launched *Sputnik*, I was thrilled to see it actually happen.

On January 31, 1958, the United States responded and launched the *Explorer 1* satellite. By this time I had done enough reading and made enough scrapbooks to understand it was scientists, physicists, chemists, and engineers doing this space work. The space program needed these professionals to build rockets and plan missions. And because a lot of the articles I read were written in Indiana, many talked about engineers from Purdue University—one of the top engineering schools in the nation.

To be honest, in fourth grade I didn't really know for sure what an engineer did. But I knew it was engineers and scientists doing this exciting work, and the engineers were the ones putting their hands on the hardware. So I decided that's what I was going to do. I was going to Purdue University, I was going to become an engineer, and I was going to work in our country's space program. I made it my life's goal.

Mrs. Laney is no longer alive, but she lived long enough to see one of her students go into space. I talked to her when I went back to Crown Point after my first flight in 1985, and I made sure she was front and center when I gave a hometown speech. I publicly thanked her for helping to spark my imagination. She said she was very proud that one of her students could do something so exciting.

After *Sputnik*, everything involved with the exploration of space started to move very quickly. In 1958 the National Aeronautics and Space Administration (NASA) was created, replacing its predecessor, the National Advisory Committee for Aeronautics (NACA). In April of 1959, NASA introduced the first US space travelers—the Mercury

Seven. Virgil "Gus" Grissom from my home state of Indiana was among them, which caused a lot of excitement where I lived. NASA called these men "astronauts," a term derived from two Greek words meaning "star sailor."

Even before the Mercury astronauts started launching, I began building my own rockets. In 1960 my cousin Keith Ross showed me an article in a magazine about Estes rockets. The article described a model kit that used little, precast, solid rocket motors. A model rocketeer could paint and decorate the rockets, but I wasn't interested in that. I just wanted to assemble and launch them. I wanted to blast them high into the sky!

The rocket motors came with an igniter. You hooked the igniter with wires to a car battery, pushed the launch button, and away the rocket flew. When the rocket had reached its maximum altitude, a parachute opened and the rocket would float back to the ground, undamaged. Hopefully.

I launched rockets and recorded details about what engines I used, how successful the rockets were, what payload they carried, and approximately how high they went. I felt like Werner von Braun. I built and launched a lot of Estes rockets. Most launches went smoothly, but some didn't go so well. My friends, sisters, and parents were my launch and landing recovery support team. They helped me search the yard, the garden, the trees, the pond, and the beanfields and cornfields for rockets—or pieces of rockets.

When Estes offered the Ranger, their biggest rocket model up to that time, I bought one and built it. Most of the rockets were about an inch in diameter, but this one was two inches, with a big payload capacity. And instead of one engine, it used a cluster of three rocket engines at the bottom. I had already been launching crickets and grasshoppers that I placed into the tops of some of my smaller rockets, and they had all survived. So I had another idea.

My younger sister, Janet, had a pet white mouse, and I decided that mouse needed to go into space. The Russians were launching dogs. The Americans were testing their capsules with chimpanzees onboard. I would launch my sister's pet mouse.

As I remember, neither my sister nor the mouse thought this was a great idea. Janet said, "You can't hurt my mouse." I told her it would be okay. "You've seen me launch grasshoppers and crickets, and they all came back fine," I assured her.

She finally accepted my plan. I was her older brother, and she trusted me. Janet also knew I was probably going to draft the mouse to "man" my rocket no matter what she said, so she might as well go along.

I carefully loaded the mouse into the payload compartment at the top of the rocket and put in some cotton balls to protect him. He looked cozy. I hooked up the connections and fired. Only two of the three engines ignited, the rocket went up askew, came down without the parachute, and crashed. I was upset. The rocket, the biggest one I had ever built, was bent and ruined.

And, later that afternoon, we had a funeral for the mouse.

To this day, whenever she has an opportunity, my sister still reminds me of her mouse's demise.

Years later, when I got married, I gave my new wife Karen an Estes kit, and she built a rocket while we were living in married student housing at Purdue University. She put in hours of effort building, painting, and detailing her rocket. It really looked nice, nicer than any I had ever built. We went to the open field by the Jet Propulsion Lab near campus where I was working as a graduate student. It was the perfect place for her to launch her first rocket and find out for herself why I was so excited about space. Once at the launch site I realized that the smallest rocket engine I had in my parts box had more thrust than the maximum engine size recommended for her model. "Wow," I thought, "this should really be cool!" I put that more powerful engine in her rocket and prepared to launch.

It really flew. In fact, it really flew clear out of sight. We never found it. It really had looked nice, but it really was gone. And she was really, really mad at me! Come to think of it, she hasn't built another rocket since.

I learned a lot from those model rockets about trajectory, center of mass, and much more. It was a very good education.

On April 12, 1961, I was finishing seventh grade when the Russians launched the first human into space—Yuri Gagarin. The Russians called him a "cosmonaut," a term derived from two Greek words meaning "universe sailor." I was upset we had been beaten again in another space first, but I knew the US would launch someone soon.

When you are thirteen years old, twenty years seems like more than a lifetime; however, it's stunning for me to look back now and realize that only twenty years to the day after Gagarin's flight, the first reusable space vehicle, the Space Shuttle, would lift off from the Kennedy Space Center. And I would be part of that program.

On May 5, 1961, NASA launched the first American, Alan Shepard, into space. At school we all gathered around a little black-and-white TV to watch. We cheered him on as his rocket lifted off the launch pad. He went up and came down. It was all over in fifteen minutes.

Al was born in 1923, just five years after the end of World War I. He was born only twenty years after the Wright brothers flew at Kitty Hawk and four years before Charles Lindbergh flew solo across the Atlantic Ocean. This man, who was born in the shadows of the first days of flight, became not only the first American in space, but also the fifth person to walk on the Moon.

I would meet this space icon several times in my life, always in group settings. He spoke to our astronaut class, and I saw him at astronaut reunions.

Al was trim and athletic-looking. He had a quick mind and alert, piercing eyes. He seemed to be pretty easygoing and liked to laugh, but at the same time I sensed respect and maybe something else from his astronaut contemporaries. As Chief, Al reportedly had run the Astronaut Office with an iron hand, and their relationship was different than just a bunch of buddies getting together to remember the good old days. We lost Al in 1998.

Twenty days after Shepard's flight, President John F. Kennedy went to the United States Congress and delivered a special message on urgent national needs. The speech included more than six thousand words. Thirty-one of those words are part of American history: "I believe that this nation should commit itself to achieving the goal, before this decade is out, of landing a man on the Moon and returning him safely to the Earth."

Now we were really on our way. Kennedy pointed out that a mission to the Moon would be "impressive" and "important." He also said it would be "difficult" and "expensive." He was right on all counts.

But, as we would learn, when the American will is coupled with the incredible science, technology, and engineering of NASA, nothing is impossible.

Gus Grissom was next into space. We were on a family vacation in Minnesota when Grissom launched on July 21, 1961. We were driving, and I knew the time for the launch was getting close. I pestered my folks to stop at a roadside café that had a television so we could watch, and they finally did. I was excited as we watched the whole fifteen-minute flight.

Fifty years to the day after Gus flew a capsule named *Liberty Bell 7* into space while I watched at a café in Minnesota, I was on the runway at the Kennedy Space Center and was one of the first to greet the astronauts when they returned from STS-135, the final US Space Shuttle mission.

Gherman Titov, a Russian, became the fourth person launched into space in August of 1961, circling the globe seventeen times. His flight proved that humans could live, work, and even sleep in space. When Yuri Gagarin died in an airplane accident in 1968, Titov was considered too much of a Soviet icon to risk losing, and his career as a cosmonaut came to an end.

In 1990 I was invited to Russia by their chapter of the Association of Space Explorers. Early on the second morning of the trip we boarded a Russian plane that took us to the launch site to see a Soyuz rocket liftoff. We were met by none other than General Titov.

I introduced myself, he started speaking, and I smelled the unmistakable aroma of alcohol. Not unlike many I met in the Soviet Union, he had already been drinking. We got on the plane, and he invited me to sit in the front section with him at his table. We talked to each other through an interpreter. He told me his house was nearby, close to the airfield, and we would be able to see it when we took off. He said something else, which the interpreter refused to translate. The interpreter kept saying, "Nyet (No)," and Titov kept replying, "Da (Yes)." Finally, the interpreter told me that General Titov was explaining that he liked to have Soviet spy satellites take pictures of his house so he could tell whose cars were parked in his driveway. I really enjoyed our flight together and the opportunity to discuss our spaceflight experiences. It was like having a live history book to talk to!

In the fall of 1962, as I started Crown Point High School, the Mercury space missions continued and Americans began orbiting the Earth. I liked high school, but I didn't like some subjects as much as others. I liked math and science; I liked history and wood shop. I didn't like French. English was something to endure.

I didn't even come close to getting straight A's, but I did well enough, and I had quite a few accelerated classes. I ended up in the top 10 percent of my class, and I made sure I took all the courses required for acceptance into the School of Engineering at Purdue.

But getting accepted to Purdue was only half the problem. The other half was making money to pay for it. I had started working to fund my college education as soon as I set my sights on attending Purdue.

Our family lived in the country, but we didn't have a farm. I did some work for my Grandpa Joe, but he normally did everything himself. That meant I had time to work on neighboring farms. I was one of the few kids in the community willing to do hard work, and it became known around the neighborhood that I wanted to make money. The farmers I worked for would call me and find out when I'd be available to bale hay or do other chores. Sometimes they would even delay cutting their hay until they knew I could help.

I didn't get paid much. Most of the time I received a dollar an hour for baling hay or for disking fields. When I got older, I was paid a dollar and a quarter or even a dollar and a half an hour. But that was the most I ever got working on farms. Mom took me to the First National Bank of Crown Point, and I opened a savings account with the money I made. I still have the bank book.

I learned a dollar earned is worth holding onto. I also learned that physical labor may be hard, but it's good for you. I enjoyed it. I enjoyed being productive.

A German farmer named Harry Gerlach and his wife Hilda lived across the road from us. They made small round bales of hay to feed their dairy herd. The baler dropped the bales on the ground, and they had to be picked up and loaded onto a wagon so they could be taken to the barn. Harry had an old Allis Chalmers tractor with a hand clutch. That made it easier for me to drive since my legs were too short to use the standard foot clutch. At ten years old, the Gerlachs hired me to drive the tractor that pulled the wagon, and Harry threw the bales onto the wagon while Hilda stacked them.

The Gerlachs were wonderful neighbors. Harry was of medium height and of relatively slender build, but he was very strong and worked like a bull all day long. He was in constant motion. He had a thick German accent that made it hard for a ten-year-old to understand his instructions. He had light brown hair and a dark tan from always being in the fields and barnyard. He rented his farm and was determinedly working toward having his own land and livestock. Harry saw the humor in things and would grin or laugh often.

Hilda was almost petite, but she was also quite strong and could do physical work like most men. She had blond hair, normally tied back in a bun or braids, and bright eyes that were only dimmed by the long hours of work that she endured with her husband pursuing their dream. She managed the house and the kids, and she spent plenty of time helping in the barn and the fields. They had three daughters and then a son, but their son died quite young from leukemia.

They mostly did dairy farming, one of the most demanding types of work that has ever been invented: milk the cows in the morning; milk the cows in the evening; and in between, grow their feed, feed them, clean the barn, and process the milk for delivery to the dairy. Seven days a week, fifty-two weeks a year!

Like most of the men in the area, Harry had fought in World War II, but his story was unique for Crown Point. He fought for the Germans. He had been captured and held as a prisoner of war at Fort Campbell, Kentucky. When the war was over, he headed back to Germany, but he didn't make it. The French detained him and made him work in their mines before they released him. When he finally reached

Germany, as soon as he could, he made a beeline back to the United States. Harry's tale was quite an eye-popping story for a country kid.

After Harry died from cancer, Hilda eventually moved back to Germany with their three daughters. When I was in Germany in 1992 training for a Space Shuttle flight, I traveled to Stuttgart and visited them.

My visit brought me right down to Earth. Hilda only knew me when I was young. She said she never thought I was that exceptional as a boy. I was a good kid, but to her I was just a neighborhood boy who worked hard. She was surprised I had become an astronaut. I think everyone who knew me as a child is amazed that I became an astronaut. Even after thirty-three years in the space business, it's still amazing to me.

Carl Deuchert was another German farmer who lived down the road from us. He was a grain farmer who also raised chickens, sold eggs, and had a Christmas tree farm. He was a very industrious man. He was serious and not very talkative. His wife's name was Lena, and they had come to the United States from Germany between World Wars I and II.

One year when Jim and I were in junior high, Mr. Deuchert hired us to whitewash his picket fence. We got more paint on the ground and ourselves than we did on the fence. Jim and I were fired from that job.

But Mr. Deuchert had more work for us. Jim and I helped Mr. and Mrs. Deuchert vaccinate their chickens. That work was done at night after the chickens had roosted. We would catch the chickens and bring them to the Deucherts, who gave them the vaccination. Jim and I made many trips to the hen house to grab the chickens by their legs and carry them upside down to Carl and Lena. The chickens were not at all happy about the whole thing.

There were also some chickens that had roosted on the branches of trees in the chicken yard, so I climbed up into the trees to grab those hens. One time I had two hens in each hand, and as I tried to get back down out of the tree, I lost my balance and had to jump to keep from falling. All four chickens flapped their wings as hard as they could as we came down. Jim said it looked like I was attempting to use the chickens to fly. It was my first experience in (under-) powered flight. All four chickens lived to receive their vaccinations. I also survived with only minor bruises.

Another summer Mr. Deuchert hired Jim and me to help him trim hundreds of Christmas trees. It was July, the sun was beating down on us, and there was no breeze in the midst of the tall pine trees. We were constantly holding the shears above our heads, and the trimmings fell into our faces. There were mosquitoes and flies buzzing all around, and we got sticky sap all over our hands. It was one of the most miserable jobs I ever had.

We trimmed the branches so the trees would be fuller, more coni-cal, and attractive by Christmas. That was a tough business, and Carl Deuchert was a tough character. He just slaved away. His hands were gnarled, and he had a finger that was partially missing from a farm-ing accident.

Harry and Hilda Gerlach and Carl and Lena Deuchert were the kind of people I grew up around, the kind of people who taught me many of life's important lessons.

As we grew up, Jim, Prancer, and I took advantage of our greater freedom to explore the fields, ponds, and creeks around our houses. Eventually equipped with bikes, we went farther afield. We rode our bikes to play sandlot baseball and football with neighborhood friends Dean and Dan Bradley, Dave Hodge, and Curt Graves. We rode down to the country store in Leroy to buy ice-cold bottles of RC Cola. We would often collect enough empty pop bottles on our way to the store to pay for the drinks.

And we explored a marsh on the land Jim's family owned. I was known for getting wet in various bodies of water. Sometimes I got wet in the pond that was near our house. Other times I got wet building dams in the creek that ran through my Grandpa Joe's field between his house and ours. But most often, I got wet in Gentlemans' marsh.

The marsh was the site of many adventures. We fished, hunted, trapped, ice skated, tobogganed, sledded, and played in the marsh, and all of that meant getting wet. In fact, when Jim's family built a new house on the land that included the marsh, I normally had a set of clothes drying in their basement to change into after our escapades.

Once Jim and I found a very old and very small rowboat on the bank of the marsh. The old wooden boat was in great need of repair. It would have been easier to burn it and build a new one, but we made up a name—"Little Aastabasta"—and tried to repair it anyway. We got tar from our parents' garages and sealed the cracks in the boat. We thought it was beautiful. And, as you can guess, we were covered with tar ourselves. I don't know how I ended up being selected to man the boat when we tested the repairs, but I did. And I got wet!

We used the Little Aastabasta to fish, to spend hot summer days on the water, and to get from place to place around the marsh. The marsh was about three feet deep in some places, but it was ten feet deep in others where ditches had been dug for drainage. Why I didn't drown is a good question, because I couldn't swim!

One day when I was baling hay on a neighbor's farm, Jim and some of our friends had one last great adventure with the Little Aastabasta in the marsh. One of them had some very large firecrackers with wicks that would burn under water. So, naturally, those firecrackers and my friends went to sea! Sitting in the boat, they packed mud around the outside of the large firecrackers, lit the wicks, and dropped the explosives over the side into the water. The firecrackers exploded under the water like depth charges dropped to destroy submarines.

Suddenly, someone noticed that the wicks of some of the firecrackers in the bottom of the boat were burning, ready to explode. Everyone abandoned ship just before the firecrackers detonated. The resulting hole in the boat was deemed irreparable, and the Little Aastabasta was given a proper "burial at sea."

There was a hill in the back of the marsh where blackberries grew —great big, juicy blackberries. Jim and I would go out to the marsh and pick them. We'd get hot, scratched up, and bitten by mosquitoes. Sometimes we'd give blackberries to Jim's mom, at least those that we didn't eat, and she'd make jam. But most of the time we'd take the blackberries down the road to Mrs. Heckel and sell them to her for eighty-five-cents per quart. Mrs. Heckel made pies and sold them, and once in a while she'd give Jim and me big pieces of blackberry pie. That was some of our best pay.

During the winters in junior high and high school, Jim and I trapped in the marsh. We set our traps and then checked them every morning and evening. Early in the morning before sunrise in the dead of winter, I rode my bike to Jim's house. I didn't want to disturb his folks, so I'd throw pebbles at his window until he woke up—if he woke up. When the light came on, if the light came on, I knew he'd be out soon, and we walked back to the marsh in the dark and checked our traps. We caught muskrat and mink. We sold muskrat furs for seventy-five cents to a dollar and fifty cents each and got up to ten bucks for a mink.

Along with working in the mills, Dad also did some subcontracting. He built additions on several houses, and he did a lot of work putting in foundations and laying concrete blocks.

When I was in junior high and high school, Dad frequently took me with him on these jobs, and he really made me work. I had to mix the mortar by hand and bring it to him in buckets, and I also had to carry the blocks to him. That's how I earned the money for my high school class ring.

The summer before my junior year, I worked as a janitor in local schools and made enough money to buy a used, small, white Honda 90 motorcycle. It was kind of a sissy-looking bike and wasn't very fast, but I needed it to get back and forth to jobs during the summer and to football practice in the fall.

JIM: *At the south end of our road there is a small, wooded area called Southeast Grove. The road through Southeast Grove has two or three hills on it, and if you take them fast enough in a car, there is a momentary feeling of weightlessness.*

When I took my first ride on the back of Jerry's Honda 90, he headed straight for that road. All of a sudden he gave the Honda a little more gas as we crested the hills. He was hooting and hollering and having a great time while my heart was in my throat.

Never trust a future astronaut on a motorcycle.

JANET ROSS RATTAZZI: *I got a ride on Jerry's motorcycle only once, but he wasn't on it. He wasn't even around.*

That school year I had a perfect attendance record. One morning things didn't go right, and I missed the bus. For some reason, we didn't have a car at home. Mom had no idea how to ride Jerry's motorcycle, but she didn't want me to ruin my perfect record. Mom got on the motorcycle with me behind her, and off we went. I don't think she ever made it out of first gear, and we fell over once. But we made it to school on time.

Most everyone in Indiana loves basketball, and almost every boy plays. Uncle Harold worked for Bell Telephone, and he was able to get us two old telephone poles. Dad and I set up those telephone poles at the end of the driveway and put a backboard and hoop on them. I practiced all summer long so I'd be ready for the eighth grade team. When I did driving layups, I had to twist so I wouldn't hit the poles. When I went back to practice at school in the fall, I kept doing the twist. Coach Fauver asked why I was twisting in such a weird way when I was doing layups. I told Coach that I had practiced all summer and that is what I had to do at home to avoid hitting the poles. He looked at me in total disbelief and said, "Well, don't do it here." After junior high I played a lot of basketball, but with my friends. We played in a hayloft in Neil Bult's barn in the fall and winter until the basketball wouldn't bounce anymore because it was too cold and stiff.

In high school I was shy and bashful. I went to very few dances, and I didn't have many girlfriends. I did not go to my high school proms. When the other kids were dancing to the Mamas and the Papas singing "California Dreaming," I was earning money and dreaming about space. I applied for admission to Purdue University during my senior year, and I was very happy to get the acceptance letter in the mailbox.

Right before graduation in the spring of 1966, I took a trip to Florida with three classmates. One of the travelers, Fred Willman, was a minister's son, and we drove Reverend Willman's station wagon. Accompanying Fred and me were Chuck Lee, who was the drum major in the band, and Tim Sendak. Tim's father would be elected Indiana Attorney General in 1968, 1972, and 1976.

I went along because I hoped that somehow I would get to go to the Kennedy Space Center. Astronauts Gene Cernan and Tom Stafford were scheduled to lift off on Gemini 9, and I wanted to see the launch. Gene had graduated from Purdue in 1956. Project Gemini, two-man flights that further demonstrated America's abilities in space, including the key capability of rendezvous, followed the Mercury missions. The third NASA program would be Apollo, which landed men on the Moon.

Cernan's Gemini flight was the third in a series of rendezvous and docking missions. For the docking test, an Agena target vehicle was launched on the top of an Atlas rocket. Later the same morning, the Gemini capsule with two men on board would blast off atop a Titan II rocket to rendezvous with the orbiting Agena.

This would be two Gemini flights in a row with Purdue graduates on board. Neil Armstrong had launched as commander of Gemini 8 in March of 1966. That crew accomplished the first docking in space, but a malfunction caused a near-fatal tumbling of the capsule/Agena combination. Armstrong was able to bring the tumbling capsule under control only after undocking from the Agena. What followed was the first emergency landing of a manned space vehicle.

The four of us left for Florida right after our baccalaureate ceremony. We took turns driving all night, and it was a long trip to Florida because the interstate highways were only partially completed.

We finally got to our destination, Daytona Beach, and thought we'd chase all the girls. But it wasn't summer yet, so the beach was empty. After knocking around Daytona Beach, we were tired, so we pulled out our sleeping bags and bedded down on the sand. It wasn't long before we realized our winter-weight sleeping bags were entirely too warm. Then it started to rain. And there may not have been any girls on the beach, but there were sand fleas all over the place.

After an hour or so, we got back in the car, tired, wet, and itching, and we started driving without a destination in mind. I said encouragingly, "Hey, we're not too far from Cape Kennedy, and there's going to be a launch later this morning. Let's drive down!" No one else had any other ideas, thank goodness!

So we went to the Cape. We got there at two o'clock in the morning and pulled off into a little park on a jetty. There was no one else around, so we collapsed on the ground and fell fast asleep.

When we woke up, there were cars and people everywhere. I'm surprised we weren't run over in the middle of the night. Everyone was excited and talking about the launches. And at 10:12 a.m., it happened! The Atlas lifted off, rapidly accelerated, and was soon out of sight. However, within minutes the news came over car radios that one of the Atlas's engines had gone out of control and the whole Atlas/Agena rocket had fallen back into the Atlantic Ocean. Since the Agena had not successfully made it into orbit, the Gemini flight with Cernan and Stafford was delayed.

I thought I knew what to expect from a launch, but I was somewhat underwhelmed by the Atlas rocket. We were at least twelve miles from the launch pad, and it took the sound a long time to reach us. The launch was cool, no doubt about it—lots of flames—but I had wanted to see the larger Titan II with the Gemini 9 capsule and men on top.

We drove home to Crown Point. I hadn't seen what I'd hoped to, but at least I had witnessed my first rocket launch from the Cape, and also experienced my first launch delay.

That summer before my freshman year at Purdue, I worked at US Steel in Gary in the blast furnace maintenance division. It was dirty, hot work, but pretty good money for a kid getting ready for college.

Crown Point had been a great place to grow up. My family was deeply rooted there. But I wanted something different. I wanted to become an engineer involved in our country's space program.

And maybe even fly.

"Look to your left and look to your right"

I entered Purdue University in the fall of 1966 when the Space Race between the United States and the Soviet Union was going full steam and the lid on college campuses nationwide was about to blow. It was a time of scientific and engineering discovery. It was a time of social upheaval. There was an America that was racing to the Moon, and there was an America that was fighting with itself about the war in Vietnam. There was an America standing up for democracy and human rights overseas, and there was an America facing civil rights issues at home.

Purdue is a Big Ten land-grant university in Indiana, with about 40,000 students on its West Lafayette campus today, but only half that when I enrolled. Covered with redbrick buildings, it was established in 1869 and has a long history of excellence in engineering, science, technology, agriculture, veterinary medicine, pharmacy, health sciences, biological sciences, education, management, hospitality and tourism management, and the liberal arts. It's known as the "Cradle of Astronauts" for the twenty-three men and women with Purdue degrees who have been selected for spaceflight.

Grandpa Joe had taken an agriculture short course at Purdue back in 1913. Uncle Don, Aunt Mary Ann, and Aunt Dorothy and her husband Bill graduated from there. But before I arrived at Purdue, all I really knew was that it had a great engineering school, that many of its graduates were working in our country's space program, and that four graduates were astronauts: Virgil "Gus" Grissom, Neil Armstrong, Gene Cernan, and Roger Chaffee.

Purdue was a perfect fit for me. There was and still is a great Midwest work ethic that permeates the entire university. I couldn't have made a better college choice. Purdue provided me with a great education and allowed me to grow personally. It gave me opportunities for leadership. And most important of all, it's where I met my wife.

Being a Purdue graduate has opened many doors in my life. In the places where I've been employed, other Purdue graduates who preceded me set the bar high. The people I worked for and with expected me to produce right away because of their previous experience with Purdue alumni. They expected Purdue people to work hard, take pride in what they did, and perform with excellence. Years later a NASA official in Human Resources at the Johnson Space Center told me that Purdue graduates were among the most dedicated and best hires in the space program. So maybe it's not surprising that through the end of the Shuttle program, about 35 percent of all US human spaceflights had a Purdue graduate on board!

Prior to arriving for my freshman year, I had to sign up for classes. One of the course requirements for freshmen was physical education, but in those days, men had the option of taking Reserve Officer Training Corps (ROTC) instead. As a land-grant university, military training has been part of Purdue since its founding.

A high school counselor had talked to me about ROTC as an option for helping me pay for college. I wasn't interested. Another counselor talked to me about going to one of the US military academies.

I had the grades, but I had no desire to do that either. Other than my father and uncles, who served in World War II, no one in my family had military experience, and I'd never known anyone with a military career. My only involvement with the military at that point had been registering for the draft when I turned eighteen and requesting a student draft deferment to attend Purdue.

Physical education was fine with me. I loved athletics. I played basketball in sixth, seventh, and eighth grades, but I was too short for high school ball. I got too many elbows on the top of my head. I played Little League, Babe Ruth, high school, Connie Mack, and American Legion baseball. I played football the last two years of high school. Unfortunately, I discovered the physical education class offered to male freshmen was not going to be any of these sports I loved. It was going to be swimming. And that was a problem.

I couldn't swim.

There was no public swimming pool in Crown Point, where I could have taken lessons. I had spent a lot of time in and around Jim Gentleman's marsh, and I got wet plenty of times. I lived half an hour from one of the most beautiful bodies of water in the world—Lake Michigan. But the few attempts I made to swim in the cold Lake Michigan water or other lakes on family vacations had been utter failures. In the water, I was an absolute rock.

In addition, the indoor swimming pool at Purdue was located on the opposite side of campus from the Circle Pines Cooperative House where I was going to live. I didn't like the idea of jumping into a pool of water and then walking across campus in the wintertime with wet hair, smelling like chlorine.

Given my options, I made a decision. I signed up for ROTC. I picked Air Force ROTC instead of Army or Navy because I knew that the Air Force was the service doing most of the military space work.

That summer and the fall of 1966 were incredibly successful for NASA. Gemini 9 with Gene Cernan and Tom Stafford (the launch I had hoped to see while visiting Florida in May) and Gemini 10 with John Young and Mike Collins flew during the summer. In September and November, Gemini 11 and 12 with Pete Conrad and Dick Gordon, and Jim Lovell and Edwin "Buzz" Aldrin, Jr., completed Project Gemini, setting the stage for the final step in the race to the Moon. The first Apollo mission was slated for February 1967. I followed all the developments on television, on the radio, and in the newspapers. I couldn't wait to be part of it.

By 1966, even TV shows were getting into space. The year before had seen the premiere of *I Dream of Jeannie*, with Barbara Eden as a genie and Larry Hagman as an astronaut and her "master." Debuting on September 8, 1966, was *Star Trek* and the voyages of the USS *Enterprise*: "Space: the final frontier. These are the voyages of the Starship *Enterprise*. Its five-year mission: to explore strange new worlds, to seek out new life and new civilizations, to boldly go where no man has gone before." I was ready for Scotty to beam me up!

I saw my life in three phases. The first was getting to Purdue, and everything I had accomplished up to then had been focused on that. I had saved enough money to pay my own way for at least the first two years. That was done. The second phase—to become an engineer—was about to begin. Phase three was to get a job in the space program.

Before the first day of classes, my parents drove me down to Purdue and dropped me off at the Circle Pines Cooperative House. We unloaded my stuff, got it all to my room, and I went outside with them and watched as they drove away. "Well," I thought, "here I go!"

I learned what I was up against on the first day of my freshman engineering class. The professor stood in the well of the big lecture hall, his voice booming throughout the room.

"Look to your left and look to your right," he said, and then paused as we glanced at the classmates beside us. "By the end of the year, only one of you will still be in engineering."

It was a statement the faculty had used to scare freshman engineering students for many years. And it worked.

Personally, I wasn't afraid of engineering. But I was afraid of failure. I was afraid of disappointing my parents, grandparents, and teachers who all believed in me. However, I was most afraid of not achieving my goal, and that would have been devastating. So I studied very hard, but just like in high school, I wasn't a straight-A student at Purdue.

I didn't study all the time. I also had some fun. I played intramural softball, mushball (sixteen-inch softball), basketball, and flag football with Circle Pines, and I was involved in other activities too. But I did not drink. I still only occasionally have an alcoholic beverage. Dad and Mom seldom drank, and drinking had a bad connotation in my mind. I didn't need anything to help me have fun, and I didn't want to start doing anything that might cause me to not achieve my goals.

That fall, with future NFL Hall of Famer Bob Griese as quarterback, Purdue had a great football team. When we beat archrival Indiana University 51 to 6 in the Old Oaken Bucket game at the end of the season, I jumped onto the field and helped tear down the north goalpost. Some of my Circle Pines house brothers and I were holding onto the same upright. We kept yelling to each other, and the other students holding onto the post thought there was a large group that "owned" that one. Slowly but surely, as we carried the post across the field and out the south end of Ross-Ade Stadium, the other students dropped off until it was just my Circle Pines brothers and me holding the post. We carried our treasure all the way across campus and back to the house. The upright from the Old Oaken Bucket game of 1966 still resides at Circle Pines today.

With the victory over Indiana, Purdue was headed to the Rose Bowl for the first time. A trip to the Rose Bowl was huge for Purdue fans. January in Indiana is cold and dark, and Pasadena is warm and colorful. I grew up in the freezing temperatures and snow watching that flower-filled parade and game every year on TV.

Dad decided that our whole family was going to California for the game. It would be our first family vacation in some time. We were to fly out of Chicago's O'Hare Field to Los Angeles, but the Chicago meteorologists predicted a huge snowstorm for the entire area and Dad got nervous. To avoid the snow, the evening before our flight the five of us packed the car and drove to O'Hare. It was about a two-hour drive when the roads were clear. And since there was absolutely no snow that night and the roads were perfectly clear, we made it in two hours. However, we could not find a hotel room near O'Hare, and we did not want to sleep in the car, so we turned around and drove home.

The next morning we got up early and drove back to O'Hare. After we arrived at the airport, the snow did finally come and delayed our flight for several hours. But we were there, and we didn't really care. We were soon on our way.

I was a college student who wanted to work in the space program. And before that moment, I had never been inside or even close to a commercial jet airplane.

We flew an American Airlines Boeing 707 on the way to California and a Boeing 737 on the way home. It was a cool ride at 30,000 feet in the air and flying so fast—nearly 500 miles per hour. I can still see myself on those flights—my nose pressed against the window. I remember flying over the Rockies, watching the headlights of cars driving on curving mountain roads below, and seeing all the city lights as we descended over Los Angeles.

During my spaceflights, as I orbited the Earth 200 miles above the Rocky Mountains, traveling at more than 17,500 miles per hour, I thought about that first airplane trip to California. I still had my nose pressed against the windows so often that fellow crewmembers teased me.

On my early Shuttle missions, I tethered my sleeping bag under the large windows on the aft flight deck. Every ninety minutes when the sun came up, the bright light woke me, and I would float to the windows and enjoy seeing whatever I could for as long as I could. I loved being up there, and I loved looking down at God's beautiful creation that we call Earth.

The view of Earth is overwhelming when you're moving at five miles per second. The best way to describe the experience is like this: think of hundreds of beautiful photos of flowers, rainbows, thunderstorms, sunrises, and features in our national parks. Then imagine that you glue them together end-to-end and run them past your eyes quickly. That's what it's like looking out the windows of the Space Shuttle. The view is a rapidly changing kaleidoscope of scenes and colors that overwhelms the senses and leaves everyone who is fortunate enough to witness it without adequate words to describe what they have seen. I never tired of the view.

The flights to California and back were actually one of the coolest parts of the Rose Bowl trip for me, but to be honest, I got a little airsick.

We left Chicago in a snowstorm, but because we expected it to be warm in California, we had all packed summer clothes. When we got off the plane, we found out it can be pretty darn cold in Los Angeles after dark, and we nearly froze while waiting for our rental car. Looking back, I guess you could say we were a bunch of country Hoosiers in the land of swimming pools and movie stars.

We stayed for three days with my mom's Aunt Ruby Shay, who lived in Tujunga. I blew a significant hole in my savings and took everyone in the family to Disneyland for a day.

On January 2, 1967, we all went to the Tournament of Roses Parade and thoroughly enjoyed seeing the beautiful flowers on the floats and the pageantry firsthand. As a student, I had purchased a ticket to the game through Purdue, and Dad was able to buy a ticket from a man at the Parade. Purdue won the game, defeating the University of Southern California 14 to 13. And I helped tear down a goalpost again.

The trip was a big deal. It was not in character for our family to spend money like this, but it was a lifetime memory. It ended up being the last family vacation that we all took together, and it also would be thirty-four years before Purdue went to the Rose Bowl again!

The second semester of my freshman year began with high expectations for our country's efforts to land a man on the Moon. Apollo 1, with Purdue graduates Gus Grissom and Roger Chaffee, along with Ed White, was scheduled to launch from the Kennedy Space Center on February 21. Apollo 1 would be Roger's first flight. He had graduated from Purdue only ten years earlier, and his wife, Martha, had been a Purdue Homecoming queen. Gus, Roger, Neil Armstrong, and Gene Cernan had all attended the 1967 Rose Bowl.

In my short nineteen years, the public attitude toward space travel had shifted 180 degrees. When I was a boy, people would compare any impossibility to going to the Moon. For example, "You might as well think you can fly to the Moon as think you can get that old car running again."

But by 1967, people believed it was only a matter of time before we reached the Moon, and NASA had an image of perfection and the ability to do the impossible. If something had "NASA" on it, people considered it to be the best and most advanced science, technology, and engineering in the world. NASA was cool, calm, and careful. NASA tested and retested. They built in redundancy. Things sometimes went wrong, like Neil Armstrong's Gemini 8 capsule spinning uncontrollably, but NASA's incredibly brave astronauts, crack engineers, and state-of-the-art technology always saved the day.

And then on January 27, 1967, all three Apollo 1 astronauts died in a cabin fire during a launch pad test. Astronauts had previously died in airplane accidents but never in a NASA space capsule. It was devastating, a great shock to the nation, and a very personal tragedy for Purdue. Some people on campus knew Gus and Roger, and there was tremendous pride in what they were doing. The Purdue community felt connected to the space program, and they still do.

I knew the manned spaceflight program would be delayed, and I wondered how that would impact our country's goal of landing a man on the Moon. The Apollo fire was the first of three low points in the history of NASA and American spaceflight, and I would be with NASA when the other two occurred. The accident didn't deter me from my dream for even an instant, and twenty months later NASA launched the first Apollo mission into orbit around Earth.

I was focused on my goal, but I still got into my share of trouble. Circle Pines was a cooperative, which meant its members were all required to do work, from cleaning to shopping to cooking. We had a black mark system to enforce the house rules. If you didn't do your assigned house chores, or if you made noise during the quiet study hours, you'd get a black mark, and there was a court system that would review them. If you were found guilty of an infraction, the court would assign a penalty, which usually entailed extra work.

During the second semester of my freshman year, I shared a study room with two fun-loving characters—Jim Adams and Stan Knafel. One night I was in our room trying to do some homework, and Jim and Stan were making fun of me, giving me a hard time, and generally making it difficult for me to concentrate. I said, "Shut up! I'm trying to study!" They ignored me and continued to have fun at my expense. Finally, one of them challenged me, "What are you going to do about it?"

I knew exactly what I was going to do about it. I had a "special," brand new can of shaving cream in my desk that seemed to be over-charged with pressurant. I had saved it for just such an occasion. I grabbed it, shook it hard, turned around, and let them have it. I nailed Jim right in the ear with a solid stream. He turned around and looked at me in disbelief. I ran down the hall with him right behind me, both of us laughing and yelling.

We did a little extra work in the house that week.

Many more shaving cream fights followed during the next three years. I still get together with these guys. We don't shoot shaving cream at each other anymore, but the stories about the battles we had are getting bigger, badder, and better with each telling. Circle Pines was a great place to live and study while I was at Purdue. I made numerous lifelong friends and have many great memories. The four groomsmen in my wedding, Jim Gentleman, Ron Tillett, Stan Knafel, and Rick Comingore, were all house brothers.

The freshman engineering courses were challenging, and I really had to work hard to learn the material. The first semester, my grades weren't great, but they were okay. The second semester, my calculus reci-tation class was at the dreaded time of 7:30 a.m. on Saturday mornings. We had lectures during the week, followed by recitation on Saturday, when we'd turn in and go over homework, have an opportunity to ask questions, and occasionally have a quiz. Homework and quizzes were part of the course grade.

Instead of going to Saturday morning class, I started going home with my Circle Pines friends on Friday nights for the weekend. We would go on dates with girls they knew from high school. I met a lot of girls, but I missed a lot of recitation classes. Consequently, at the end of my second semester, I got a D in the five-hour calculus class. That got my attention. I had to retake that five-hour class the next semes-ter along with another four-hour calculus class. That made nine hours of calculus in one semester. That's a lot of calculus, even for a guy who liked math.

To me, with that D, I had failed, and I was angry with myself. I was the first one from my immediate family to go to college, and I wasn't going to allow this to happen again.

Even with the D, my grades were good enough that I was offered a US Air Force ROTC scholarship. During my freshman year, I actually found ROTC interesting. I didn't like the marching and spit and polish, but some of the classroom work was military history, and I found that fascinating. I've always loved history, and I enjoyed learning about the military and the Air Force. For the next three years, the ROTC scholarship would cover all of my tuition and provide some additional money for books.

Accepting the scholarship meant I would have a four-year military commitment after graduation, but I accepted it for two reasons. First, I knew both my sisters would soon be enrolling at Purdue, and I was trying to help my parents by paying my college expenses myself. Second, an idea was coalescing in my head. I suppose it had always been there, but now it was becoming more clear. Instead of just being an engineer working in the space program, I wanted to be an astronaut. The Air Force trained pilots. The best pilots became test pilots, and the best test pilots could become astronauts. Accepting the ROTC scholarship and becoming an Air Force pilot provided me with a potential path to becoming an astronaut.

At the end of our freshman year, engineering students—those who were still around to look to the left and to the right—had to select a specific major. I had considered going into aeronautical engineering because I knew aeronautical engineers were building rockets and spacecraft; however, by that time, aeronautical engineers on the West Coast were without professional jobs and were literally bagging groceries. The aerospace job market was totally flat. I didn't want to bag groceries, so I took a second look and decided to go into mechanical engineering, which is the root of aeronautical engineering in the first place. I concentrated on propulsion and thermodynamics in my undergraduate work, and I did my master's work in propulsion.

I've never been disappointed with the decision. Most of the work I did in the Air Force and at NASA was more related to aeronautical engineering than mechanical engineering, but mechanical engineering gave me all the basics I needed to do the work.

In the summer of 1967, after I accepted the scholarship, the war in Vietnam was picking up. I expected that I would complete my education at Purdue, go into pilot training, and then be sent to Vietnam. That's what was happening to ROTC pilot candidates at that time.

I certainly did not want to go to Vietnam. I didn't want any part of killing people, much less of being shot at. I was hoping that we would find an end to the conflict before I completed pilot training, or I'd be stationed somewhere else. But if Vietnam was where my country needed me, I would go.

Like many students, I had mixed feelings about the Vietnam War. I didn't like that our country was losing young men in Vietnam. But at that time, I thought the war in Vietnam was the right thing to do to stop communism before it overtook more of the world, and I believed in trying to free people so they could live under their own rule. In retrospect, the war in Vietnam was not handled well, and it cost the lives of way too many young Americans.

I understood some of the protests, but I was too much country and apple pie to agree with everything the antiwar groups were doing. Like most other students at Purdue, I stayed focused on school.

In the beginning of my sophomore year, I expected to qualify as a pilot training candidate with the Air Force. I took a physical examination that included a vision test and passed with flying colors. My eyesight was better than normal, 20/15, but I had to undergo a more comprehensive, dilated eye exam to complete the physical. The ROTC unit set up an appointment with an ophthalmologist about an hour away at Bunker Hill Air Force Base, near Peru, Indiana. Bunker Hill would later be renamed in honor of Gus Grissom.

Unfortunately, I overslept and missed the appointment, the only time I can remember doing so in my life. I was upset with myself, but it wasn't a problem. The Air Force made another appointment for me, and this time it was right on campus and I was on time.

During the examination, I learned the ophthalmologist had been drafted into the service, and he was not at all happy about it. He complained about being in the Air Force the whole time I was there. He told me my eyes were fine; however, then he dilated them, did some tests, and told me that someday, in the years to come, I would need glasses.

And since at some point in the future I was going to need glasses, he disqualified me for pilot training.

I felt like someone had punched me in the gut!

I know now that I should have appealed the doctor's decision, and I probably would have won, but I was a sophomore in college and didn't know any better. As it turned out, I didn't need glasses until I was in my forties and had already flown in space.

Up to that point, everything had gone perfectly. Life is easy when things are going well. I found that I had to reach down inside and find out what I was really made of. It was the first real setback in fulfilling my dreams, and I hadn't seen it coming, nor did I know there would be two more big setbacks ahead.

I was upset, but I had to accept the fact that I wouldn't be a pilot. I'd just be an engineer in the Air Force. Maybe I wasn't going to get to fly into space. Maybe being an engineer was God's plan for me. Even if I couldn't be an astronaut, I could still work in the space program. I kept my focus.

Then something else happened, and this is the way life often goes. It's what people mean when they say, "When God closes a door, He opens another." It's true. But if you're too busy pouting, complaining, and thinking the world is against you, you'll miss that open door and walk smack into a wall! I believed if I worked hard and kept my sights on my goals, God would show me the open door. And He did.

NASA accepts astronauts in classes or groups. Group 1 in 1959 was the Mercury Seven. Group 2 was accepted in 1962, and that class included Neil Armstrong. Group 3 in 1963 incorporated Gene Cernan and Roger Chaffee, who had known one another at Purdue. There was a common factor among these groups: all of the astronauts were top test pilots.

At the end of my junior year in high school, NASA selected Group 4, and they were scientists—six scientists whose academic backgrounds were needed in the broadening space program. One of them, Harrison Schmitt, a geologist, would fly with Gene Cernan on the Apollo 17 mission and collect rocks on the lunar surface. In 1966, NASA selected Group 5—again all test pilots.

However, in the fall of 1967, about the same time I was being rejected as a pilot candidate by the Air Force, NASA announced the next group. The astronauts selected in Group 6 were scientists and, for the first time, engineers. Seven members of this class would eventually fly as mission specialists on the Space Shuttle, including Joe Allen, a fellow Indiana native, and Story Musgrave, who in 1996, at the age of sixty-one, became one of the oldest people to fly in space.

Encouraged that I still had a chance to become an astronaut, I continued studying and remained active on campus. At Circle Pines I served as the assistant treasurer, then the treasurer, the pledge class trainer, the membership chairman, and a member of the "judicial" standards committee (when I wasn't involved in shaving cream fights). I ran and was selected for the Purdue Sophomore Class Council.

In the middle of my sophomore year, I was "tapped" for Tomahawk, a Big Ten activities honorary for independent, non-Greek students. In the freezing early morning hours of February 21, 1968, the new pledges were awakened and brought by active Tomahawk members to the grave of university benefactor John Purdue on the Memorial Mall. A short ceremony welcomed us as the new pledge class.

I really hadn't known all that much about Tomahawk before I became a pledge, but it was fun, and I was really surprised when I was elected pledge class president. I enjoyed Tomahawk. The pledges were a nice bunch of folks, one of them in particular.

Her name was Karen Pearson, and she was a sophomore from Sheridan, Indiana. Her father, Morris Pearson, was a farmer, and he was in the sheep business. Karen and her sisters, Linda and Sandra, were very active in 4-H growing up, and at Purdue, Karen was studying vocational home economics education.

I noticed her right away. There was a quick attraction, at least on my part. I liked her personality, her good looks, her smile, her dimple. I had always been attracted to petite girls, and she was very petite. Our first date was on March 16 to watch the Indiana High School Basketball State Finals Championship game. Only in Indiana do college students spend their first date watching high school basketball!

———————————————●———————————————

KAREN ROSS: *Jerry didn't ask me out until our Toma-hawk pledge period was almost over. On our first date we watched the basketball game on TV at Circle Pines. Then we walked down the street to a little park, and we sat on the swings and talked. That might not sound exciting, but it was a perfect evening.*

I thought Jerry was smart and cute. That was my second impression of him. I had actually noticed his picture the year before in the Purdue yearbook, the Debris. *I liked Jerry's eyes and smile.*

While we were sitting on those swings, Jerry told me about his room in the basement of his house in Crown Point, about lying on his bed in the summer, tuning in distant stations on the radio. And he told me he was interested in the space program.

We were a farm family, and I loved growing up in the country. I hoped I would marry an agriculture student and live on an Indiana farm not too far from my family. I considered doing other things, but I always came back to wanting to marry a farmer and raise a family in Indiana.

I remember the moment when Jerry told me his major was engineering. My heart sank. I didn't want to get in-volved with someone who wasn't in agriculture!

I went home for Easter weekend. While I was driving home from the sunrise service with my sister Sandra, I had an accident at a crossroad a half a mile from home. San-

dra and I were both badly injured. We were in the hospital for two weeks, and I was unable to finish that semester at Purdue.

After I got home from the hospital, Jerry's Circle Pines house brother Stan Knafel brought Jerry to my house to see me, and they met my family. Jerry and I lived only about 120 miles apart, but my family thought he talked like a Chicagoan, and Jerry thought my family had southern accents.

After final exams, Jerry came to see me again. He took me for a drive in the country, and we talked about the fact that I would have to go to school an extra semester to finish my degree. Jerry told me he had found out about an Air Force program called Educational Delay that would allow him to stay at Purdue for a master's degree before entering active duty. If he did that, he could stay at Purdue while I was finishing school. I was surprised that he was thinking that far ahead about us.

———————————————————•———————————————————

Karen and I lived more than a two-hour drive apart, so we only saw each other a few times that summer. But we wrote, and after her cast was removed, I brought Karen to my house where she met my parents and sisters and stayed for a few days.

For a while Karen wasn't sure she wanted to continue seeing me because I was in engineering and had a four-year Air Force commitment. She loved living in the country in Indiana, and I didn't know where my career might lead.

———————————————————•———————————————————

KAREN: *Jerry wasn't going to be a farmer, but he was very nice, and he was serious about school. He always had a big smile, and his family was a lot like mine. We both grew up in the country and had similar values. We were*

*even both Methodists. Jerry gave me his Circle Pines pin
that fall semester of our junior year, and we looked at wedding rings in January 1969, during semester break. Soon
we were talking about wedding dates and children's names.*

———————————————●———————————————

A lot of college students get out of the habit of attending church while
they're in school, but throughout college, I attended First Methodist
Church in West Lafayette. It was a half-mile walk from Circle Pines.
Karen also attended First Methodist, but we never met there. During
the fall of 1968, we started attending Sunday services together, which
continued after we were married.

It's always been important for me to go to church every Sunday.
I have all my life, and I still do. I think it is important to support the
wider church and Christian community. I feel like church and Sunday school are my weekly opportunities to join with the community
of faith to celebrate God's love and to "recharge" my Christian battery.

Demonstrating the Christian example of regular worship for my
children and grandchildren is one of the most important things I can
do. Many Christians today don't regularly attend worship services. I
believe that is too bad—for them, for their children and grandchildren,
for the rest of the Christian community, and for the whole world.

My grades went up significantly after Karen and I started dating.
We would want to get together on Friday and Saturday, and I'd have to
redouble my efforts to get things done so I'd have time to spend with her.

———————————————●———————————————

JIM GENTLEMAN: *When I got to Purdue the year
after Jerry, I also lived in Circle Pines. During the second
semester of my sophomore year, Jerry was my roommate. It
seemed I followed Jerry everywhere but into space.*

*When we were kids and Jerry wanted me to wake up
and check our traps, he would throw pebbles against my*

window. Well, one afternoon at Purdue I heard pebbles pinging against our window at Circles Pines. I looked outside and it was Jerry's girlfriend, Karen. She was signaling to Jerry to come outside.

I thought, "Oh, no! Here we go with the pebbles again..."

Karen and I were young, but we were ready to get married. I felt mature and wanted to get on with life. Once I make a decision, I don't keep rolling things over in my mind. There has never been an important decision I have made that I would reverse. I don't procrastinate. I met a girl. I liked the girl, a lot. We wanted to get married.

We applied for married student housing even before we were engaged. We could get into the housing in the spring semester of 1970, our senior year, so we decided we would get married in January 1970, during semester break. We reasoned we could be careful with our money, saying, "Two can live as cheaply as one."

Karen and I were engaged May 3, 1969, at the end of our junior year. We went to the Circle Pines house dance, but it was really crowded and noisy, so I took her for a drive over to a big park nearby in Lafayette—Columbian Park. I asked her if she'd marry me. I was trying to surprise her a little, but she saw it coming a mile away.

KAREN: *I knew Jerry and I were going to be engaged. We had picked out rings, and I guessed he would ask me to marry him the night of the house dance. I was restless the afternoon before the dance. I didn't want to study. It was a beautiful day, and I persuaded a friend to walk with me to Triple XXX, a landmark restaurant down the hill from campus. I had ridden past the iconic orange and black striped building many times but had never eaten there. I enjoyed a hotdog and a frosty mug of Triple XXX root beer. That*

afternoon is a sentimental memory. I can remember feel-
ing that my life was about to take a wonderful and excit-
ing turn and would never be the same. And that is the only
time I have ever been to Triple XXX.

During the dance, Jerry took me to Columbian Park.
After a short walk we sat down on a park bench, and he
proposed.

I said, "Yes!"

The summer before we got married, I wanted to earn as much money
as possible. I worked eight-hour days at US Steel and then four hours
weekday evenings and on weekends as a bellhop at a Holiday Inn in
Merrillville, just north of Crown Point. Toward the end of the summer I quit the hotel job because I was given the opportunity to work
twelve hours a day, seven days a week at the mill. It was great money.

That was a big summer. On July 16, 1969, Apollo 11 launched toward the Moon to attempt the first manned lunar landing. I read everything I could get my hands on about the mission. Any time there
was information about the mission on TV, and I wasn't working, I was
there. I didn't care if the coverage was just a shot of Mission Control in
Houston with no one talking. I loved what they were doing, how they
were doing it, the suspense, and the technology. It all just fascinated
me. On July 20, Neil Armstrong and Edwin "Buzz" Aldrin, Jr., landed
on the Moon while Michael Collins orbited the Moon in the Apollo
capsule. Karen was visiting at my parent's house, and she sat on my
lap that night as the world watched Neil and Buzz step onto the lunar
surface. It was incredible to listen to Neil and watch those first steps.
It was thrilling to realize that it was actually happening.

The rest of that year seemed to fly by as Karen and her mother
prepared for our wedding. It snowed heavily during semester break
in January 1970, and the afternoon of our wedding rehearsal I got
stuck in a snow drift on the way to Karen's house. Fortunately, Ron
Johnson, a Circle Pines house brother and neighbor of Karen, came

by with a tractor, pulled my car out, and drove ahead of me to clear a path to Karen's.

Karen and I were married in a beautiful candlelit ceremony in her home church, Boxley United Methodist, out in the country on a snowy, windy Sunday evening. We had a big honeymoon—two nights in Indianapolis. Then we drove to our new home in married student housing on Nimitz Drive at Purdue and settled in.

Married student housing wasn't exactly palatial. Our apartment had a bedroom, bath, and one room that served as a living area, dining room, and kitchen. Rent was $110 a month. But it was wonderful, and we were happy.

I graduated in May of 1970 with my bachelor's degree in mechanical engineering and was commissioned as a Second Lieutenant in the US Air Force. I was staying at Purdue for my master's degree, and I already was enrolled in the industrial administration program in Purdue's Krannert School of Management. But Dr. Cecil Warner, who was my advisor for my mechanical engineering senior design project, offered me a half-time research assistantship to work with him on a US Navy-funded ramjet propulsion research project. The research opportunity and the money from the assistantship easily convinced me to change my plans from the business school back to mechanical engineering.

I believe Dr. Warner's offer was one of those providential occurrences, another of a whole series of events in my life when God possibly intervened and helped me progress toward my goal. The research assistantship came unexpectedly, and it helped change the rest of my life. It redirected my path and allowed me to get a master of science degree in mechanical engineering. That advanced technical degree opened the doors for me to obtain the types of assignments and experiences in the Air Force that permitted me to successfully apply to be an astronaut.

That summer before I started my master's program, I attempted to get work doing highway construction. I showed up early in the morning at the foreman's shack seven days in a row before I was finally hired. After watching me work for three days, the foreman told me he would have jobs the following Monday for three of my friends who had also been there looking for work. We worked highway construction on I-65 just north of Frankfort, Indiana, for half the summer. I did all kinds of jobs. I helped set concrete forms and took them down after the concrete had cured. I moved the forms and set them up again. I got pretty good with a sledgehammer. By the end of the summer, I could drive stakes into the ground by keeping the sledge rotating in a circle and striking the stake every time the hammer came around on the down swing.

The second half of the summer, I worked on rebuilding part of US 30 in north central Indiana near Warsaw. Ron Tillett and I found a place to rent in the Warsaw area. We stayed there during the week, drove a couple of hours on Saturday afternoon to get home, and went back to Warsaw on Sunday night. At the end of the summer I was in a couple of weddings, including Ron's. My other Circle Pines house brothers didn't recognize me when they saw an extremely tanned, sun-bleached blond guy standing in the line of groomsmen.

Karen and I were doing pretty well living as cheaply as one, and then Karen became pregnant. She finished her bachelor's degree in January of 1971, and our first child was due in March. It was exciting. I wanted a family right away. For whatever reasons, many of my school classmates had older parents who didn't participate in their children's activities. My folks were young when I was born, and they were active in my life. Dad coached my teams, and Mom brought cookies and cupcakes to school and helped the teachers with class activities. I wanted to be young like my parents when my children were growing up so I could be involved in their activities.

With a baby coming, it became harder to make ends meet. As students, we didn't have health insurance, so we paid the medical expenses ourselves. Fortunately, medical costs weren't then what they are now. When she was still in school, Karen's good grades had qualified her for college student loans. Before we got married, Karen and I decided she would take out a loan to give us a little margin, and with a baby on the way that loan came in handy.

Our cute little daughter, Amy Jo, was born on March 30, 1971.

It was soon clear that three couldn't live as cheaply as one. It was a big decision for Karen and me to buy a Chef Boyardee pizza mix, a six-pack of Pepsi, and a Jiffy cake mix so we could have friends over for supper and cards.

While working on my master's degree, I met active duty Air Force officers who were working on advanced engineering degrees at Purdue. In the fall of 1971, during my last semester, I told a couple of them I would like to be stationed at Wright-Patterson Air Force Base in Dayton, Ohio. Wright-Patterson wasn't far from Karen's family and mine—and that was important because Amy was eight months old, and we already had another child on the way. We weren't wasting time! We wanted to be close enough to Indiana that our children would know their grandparents, aunts, uncles, and cousins.

One of the officers I talked with pulled a Wright-Patterson telephone book from his desk drawer. Since my research project at Purdue was in ramjet propulsion, he gave me the telephone number of Blaise Durante, the Executive Officer at the Aero Propulsion Laboratory at Wright-Patterson.

Just before Thanksgiving break in 1971, I called Blaise. I told him about my graduate studies and that the Air Force had assigned me to Systems Command. I told him I wanted to work in the Aero Propulsion Laboratory at Wright-Patterson, and he said he would look into it.

After that phone call we went to Karen's folks' for Thanksgiving. Blaise actually called me back within a few hours ready to offer me a job, but since this was before answering machines and cell phones, I didn't find out about the job until he reached me when I returned to campus on Monday.

I received my master's degree from Purdue University in January of 1972 and looked forward to beginning my Air Force career.

I have great memories of those days at Purdue. I wouldn't trade them for anything. We had great friends and great times, and we received great educations. It had been an eventful five and a half years since I first arrived on campus and looked to my left and looked to my right, wondering if I was one of the guys who would make it.

THREE

"My daddy is an astronaut..."

On a snowy January day in 1972, Karen and I took a day trip to Wright-Patterson Air Force Base near Dayton, Ohio, to visit the laboratory where I would be working and to make arrangements to live in military housing.

That morning we learned there was no base housing available for us, so we had only the afternoon to find something else. After a whirl-wind few hours spent looking at apartments and duplexes, we rented an older house. Then I did something I'd never done before and I've never done since. I asked my folks if we could borrow $2,000 to get a sofa, a bedroom set, a small dinette set, and a washer and a dryer for all of those diapers. They loaned us the money and said they were very happy they could do it.

My dad worked very hard, and he never had a lot of money. I knew that. What I didn't know until many years later was that Dad had to take a loan against an insurance policy so we could have that $2,000 for some furniture. I paid the money back to my parents as soon as I could. I don't like to owe anyone, and I was especially concerned about getting the money back to my dad—even without knowing what he had done. My Second Lieutenant salary wasn't great, but it was enough to let me start paying him back. I also umpired softball games and taught a night course in physics at Sinclair Community College in Dayton so I could pay back the money quicker.

In February I drove my pregnant wife, Karen, and my snuggly ten-month-old daughter, Amy, to Fairborn, Ohio. Karen set up house and prepared for the arrival of our second child while I started my US Air Force career.

Wright-Patterson Air Force Base has a long history of flight test dating back to the Wright brothers. Orville and Wilbur Wright flew their flying machines in an area called Huffman Prairie, which is now part of the base. Among the airfields near Dayton, one was named for Wilbur Wright and another for Frank Stuart Patterson, the son and nephew of the Dayton founders of National Cash Register Company. Frank Patterson was killed in a test flight in 1918. The two fields were merged in 1948 not long after I was born.

At Wright-Patterson I worked as a ramjet engine project engineer in the Aero Propulsion Laboratory. I was put into a group with three other engineers, two captains and one civil servant. Our job was to provide computer programming support to the entire Ramjet Engine Division. I had very little programming background because that technology was still fairly new; however, I learned quickly, and it was a good thing because in less than two years, all three of the people I worked with were gone, and I was the only one left to provide computer support for Ramjets.

That year, 1972, was a very significant one in my life. I received my master's degree and entered active duty with the Air Force. Our handsome son, Scott Lee, was born on April 27. The Air Force awarded a contract to build the B-1A bomber, which was going to play an important role in my career, and NASA issued contracts for the design and construction of the Space Shuttle.

Astronauts for the Shuttle program would be classified as either pilots or missions specialists. Mission specialists would primarily be scientists and engineers, and they would perform the spacewalks, operate the robotic arm, and conduct experiments—some of the most exciting work, as far as I was concerned.

At about the same time, another incredible opportunity was being created. The Air Force and Navy had always restricted their test pilot schools to pilots. But in response to the increasing complexity of new aircraft and their electronic systems, the schools were preparing to add training programs for engineers. Engineering graduates would be called "flight test engineers," and they would work with the top military flight test programs—the same programs where NASA found many of its test pilot astronauts.

The Navy started its Flight Test Engineer program first and accepted people from other branches of service. I eagerly prepared to apply.

──────────────── ● ────────────────

KAREN ROSS: *I tried to talk to Jerry several times about our future: "What do you think you want to do? What do you think our life is going to be like?" And Jerry would answer, "I don't know. I don't want to close any doors; we will just have to wait to see what happens."*

One day not long after Scott was born, Jerry sat me down to talk. He told me about a new vehicle NASA was working on called the Space Shuttle. I had never heard of it, but he knew all about it. Jerry told me some of the astronauts on the Shuttle would be mission specialists, not pilots. He wanted to go to Test Pilot School and become a Flight Test Engineer so he would be in a better position to apply when NASA needed astronauts for the Space Shuttle program.

Wow! I knew Jerry was really interested in the space program, but until that moment I didn't know he wanted to fly in space. The thought of that was a little unbelievable; Houston and NASA seemed like a different world.

I thought, "NASA selects so few astronauts, and the people they choose are so extraordinary." It was wonderful that he wanted to do this. Jerry was special to me. But I was afraid he was aiming too high and setting himself up for disappointment.

It surprised Karen that I wanted to fly in space. But to me, it was just a logical extension of where I had been going all along. The opportunities had just not presented themselves before and now—they were all coming together at once. I always wanted to be in the space business, and if you're going to be in the space business, the best place to be is on the top of the rocket.

As I prepared to apply, I discovered that I didn't have enough time on active duty to qualify for the Navy's program. And there was another problem, which was more difficult. To qualify for the US Navy Test Pilot School, I had to be able to swim. I still couldn't swim, not a stroke. My choice of ROTC over swimming at Purdue had come back to haunt me.

After doing computer programming for a year, I was given an additional job in the Ramjet Engine Division, this time with hardware. A ramjet is an air-breathing engine that uses its high velocity to compress the air in the process of producing thrust. My job was to conduct tests at Holloman Air Force Base in New Mexico, north of El Paso, Texas. Out there on the salt flats, we put a ramjet-powered missile onto a rocket-powered sled and shot it down a seven-mile track to test the ramjet engine. The sled and attached missile achieved Mach 2.7, or almost three times the speed of sound in about ten seconds. That was really impressive!

Many of my coworkers had learned about my high-flying interests, so some of the technicians who worked at the track offered to put a saddle on the sled for me. It was tempting. I wanted to ride those rockets!

By this time the Flight Test Engineer course at the US Air Force Test Pilot School was up and running, and I didn't have to swim to be accepted. I applied, although as a very junior officer, I didn't think I had a chance of being selected on my first try. But I thought, if I was persistent, I might eventually have a chance. One thing I have learned in my life is you don't get anywhere without believing in yourself and taking some chances.

I did well on both jobs in the Ramjet Engine Division, and my supervisor Bill Supp nominated me for awards. I received two achievement awards and was nearly selected as the Air Force Systems Command Junior Officer of the Year. I was named the Aero Propulsion Laboratory's Junior Officer of the Year. My career was going well, but events that could derail my plans were brewing.

A command change was about to take place in the Aero Propulsion Laboratory, and I was asked to be the Executive Officer for the incoming commander—contingent on his approval. I told the departing Commander, Colonel Walter Moe, that I would be honored to serve as the Executive Officer.

When the new Commander, Colonel Daniel Cheatham, arrived, I told him I had applied for the Test Pilot School's Flight Test Engineer course at Edwards Air Force Base in California.

"If I'm going to accept you as my Executive Officer," he said, "I want you to commit to stay with me for at least one year. If you are accepted for Test Pilot School, I will expect you to decline. But I'll do my best to explain the circumstances to the Test Pilot School Commandant and ask him that your commitment to me not damage your chances for future selection."

I gave him my commitment.

I thought being the Executive Officer would be good experience. I would get to manage people and have close working relationships with the management team of the Aero Propulsion Laboratory. And frankly, I thought it might be a couple of years before I even had a chance of being selected for Flight Test Engineer training.

But I was wrong! I was selected as an alternate for the next class, and when someone dropped out, I received orders to report to Edwards Air Force Base. I wanted to go to Edwards! It was the next key step toward the goal I had been working on since I was ten years old. This was it.

I talked to Colonel Cheatham. I told him going to Test Pilot School was something I really wanted to do, and it was an incredible opportunity that might not come my way again. "You made a commitment," Colonel Cheatham said. He didn't have to say anything more.

I believe a person has to keep his word, no matter how difficult the situation. But this was very hard. I was afraid that I might never be selected again. The odds were stacked against me being selected the first time, and there were plenty of other qualified people who would jump at the assignment. Why mess with someone who has refused a once-in-a-lifetime opportunity?

Setback number two.

Working as Executive Officer that year was the great learning experience I had expected, and I gained leadership skills that have benefited me throughout my career. And my Commander was true to his word. Colonel Cheatham called Colonel Joseph A. Guthrie, Jr., the Commandant of the Test Pilot School at Edwards. He explained the situation and asked that turning down the assignment not be held against me.

The next time the Air Force selected a class for Test Pilot School, I was not picked as an alternate; I was on the prime list! Being selected for Flight Test Engineer training was the ultimate opportunity to fly in high-performance jet airplanes, and thinking about where it might lead was pretty cool.

But Karen had some concerns about going to Edwards.

KAREN: *I did not go easily. Jerry always volunteered to help Daddy on the farm whenever we went home. That spring Jerry was on the tractor disking, and I was riding along. Everything was green and the air was soft and warm. Before it was too late, I took a deep breath and appealed to Jerry, "You can turn this down and complete your four-year commitment to the Air Force at Wright-Pat, and we can come home." I really, really wanted to live in Indiana close to family, and if Jerry went to Test*

Pilot School, he would extend his Air Force commitment four more years.

This was a watershed moment in our lives. I asked him if he was really sure this was what he wanted to do. I told him he could be an engineer in Indiana.

Jerry said he had thought and prayed about it, and he had no doubts. I actually knew that was going to be his answer, but I had to ask. We were going to California.

I had been afraid Jerry was setting himself up for disappointment. It didn't turn out that way. And from the moment Jerry was selected for Test Pilot School, I have felt like I'm holding onto the tail of a comet.

———————————————●———————————————

I listened very carefully to what Karen told me. I understood her feelings. I told her that I liked Indiana, too, but they don't launch rockets in Indiana.

In the summer of 1975, we left the Midwest for the high desert of California. And as a result, for the next twenty years we spent most of our vacations taking our kids back to see their grandparents, aunts, uncles, and cousins in Indiana. We went home for many Christmases, some special occasions, and for an extended stay almost every summer.

When we got married between semesters at Purdue, Karen and I only had a two-day honeymoon, so we planned our drive to California as our belated wedding trip. Karen had never been west of Missouri, so after the moving van left Fairborn, we went to see her parents and then further north to see mine. I helped my dad frame up a garage, and then Karen and I took off by car on our first real trip together while our kids stayed with my folks.

We stopped at beautiful places we'd never seen: the Badlands, Mount Rushmore, Yellowstone, the Tetons, the Grand Canyon, Bryce Canyon, and Zion National Park—all places I would one day view from space.

Our last stop was Las Vegas. It was terribly hot. When we stepped outside at ten o'clock at night, the furnace-like wind took our breath away. Our motel room was not very cool, I couldn't sleep, and our car didn't have air conditioning, so we decided not to spend the night walking the Strip. Instead, we drove across the desert during the cooler night hours to Edwards in south-central California.

How could I sleep anyway? I was getting more and more excited the closer we got to Edwards. The base is named for test pilot Glen Edwards, who died along with a crew of five in a crash in 1948. The dry lake beds there provide smooth, extra-long natural runways, making it an ideal place for flight testing. Almost all US Air Force aircraft since the 1950s have been tested at Edwards. This is where Chuck Yeager exceeded the speed of sound for the first time in 1947. This is where a fellow Purdue graduate named Iven Kincheloe, nicknamed "America's No. 1 Spaceman," flew the Bell X-2 to a height of 126,200 feet in 1956. This is where Neil Armstrong flew the X-15, which later became the first airplane to enter space. Edwards was where flight legends were born, from Pancho Barnes' Happy Bottom Riding Club to the X-15A-2 that flew at Mach 6.72—still the fastest speed for any airplane.

Edwards was where the action was, and I was going to be part of it. I couldn't wait!

We drove all night and arrived at Edwards as the sun was rising. We entered through the north gate and drove about ten miles before we saw anything—and I mean anything. It seemed to get drier and browner and uglier every inch of the way. I could see the look on Karen's face. This was an Indiana farm girl used to rich, black soil and lush, green fields. I knew she was thinking, "Where in the world is he taking me?"

Karen and I checked into temporary base housing called Desert Villa, which was really basic. It was like married student housing at Purdue. The desert heat was miserable, and much to our disbelief, we didn't have any air conditioning. We searched, and we couldn't find a unit anywhere. Finally, after a couple of days, someone clued me in on the cooling system. In the very dry climate of the Mojave Desert, evaporation units called "swamp coolers" are used to cool houses. The swamp cooler was on the roof, and it worked just fine once we flipped the inconspicuous switch on the wall. Some hotshot engineer I was! Clearly, I had a lot to learn at Edwards.

Dad and Mom flew with Amy and Scott out to California. My folks took one look at Edwards and thought it was a Godforsaken place. Karen said my dad wondered what I was doing taking my family out there. But after our family had lived at Edwards for a while, we grew to enjoy it very much.

I was anxious to start flying as soon as possible. The Test Pilot School class that was nearly finished with its training was conducting a "Dynamic Demo" flight, and fellow classmate Doug Picha and I signed up. They took us along for a ride in a KC-135 refueling tanker, a Boeing 707 look-alike. As the "Dynamic Demo" flight description indicated, they were performing all kinds of unusual maneuvers with that airplane, and Doug and I were riding along like raw hamburger in the cargo area. This was our first experience in an airplane doing flight test maneuvers, and soon neither Doug nor I felt very good. When we landed, I went home and lay down on the bed. I was thinking, "I'm miserable; I'm not sure I should have applied for this. If I am going to feel like this all the time, maybe I can't do this. Maybe I should quit?"

Karen found me. She told me I looked green. Until that day she thought that was just an expression, but there I was, a shade of pale pea green, which looks great on vegetables but is not what you want to see on the face of a flight test engineer hopeful.

"You said when you got back from flying, we were going to go to Lancaster," Karen said. We had planned a trip to Lancaster, the nearest town of any size, that afternoon to get things we needed to settle into our new home.

Karen was in no mood to be sympathetic. There we were at Edwards because Test Pilot School was what I wanted, and now I had suggested I might quit. "You brought us here, and we need things," she said. "Now get up and let's go."

There's nothing like a wife to straighten up a husband. I got my green, sorry self up, and we were off to Lancaster.

At Test Pilot School, we flew in the morning and had classes in the afternoon. The concept was for the flight test engineers to have the same academic training as the test pilots and to work with them to plan the flights. We then flew with the pilots to conduct the tests and collect the data. Together we collated the data, analyzed it, and wrote a report on our findings. Having a flight test engineer in the cockpit allows a pilot to concentrate on flying precisely while the engineer collects the data. This concept has proven to be a very effective way to evaluate airplanes and their systems.

My first flight in the official school curriculum was in a McDonnell Douglas RF-4C Phantom II. It's a supersonic fighter/bomber and a really cool airplane. My pilot was "Rotten Ralph" Luczak. I never knew where Ralph's nickname originated, but many flyers' monikers were "earned" under conditions that the awardees would rather forget!

Ralph was a happy-go-lucky kind of guy who frequently had a joke to tell. Some of them were funny. He had been what we in the Air Force called a "heavy driver." He didn't fly fighters. He flew big heavy tankers. Sometimes fighter jocks don't fraternize with "heavy drivers," but Rotten Ralph was always included in the party.

As soon as Rotten Ralph and I got off the ground on that first flight, a big puff of white smoke came up behind Ralph's left shoulder. I told him about the smoke, we switched our masks to 100 percent oxygen to protect our breathing, declared an emergency to the control tower, and came back down for an emergency landing. What a way to be introduced to the world of flight testing! Fortunately, neither Ralph nor I had "rotten" luck that day.

Flight testing is dangerous work, and people were killed during my time at Edwards. Whenever we came in on an emergency landing, fire trucks and ambulances were waiting for us. That was all part of the business.

When you're in the flying game, I think most flyers have a mentality that what they are doing is dangerous, but nothing bad will happen to them. The unthinkable might happen to someone else, but not to you. You can't focus on the risk. You have to focus on what you're doing and on doing it right, because many times you're just a split second away from disaster.

---•---

KAREN: *I was aware that what Jerry was doing was inherently dangerous. There were planes that crashed and people who died while we were at Edwards, but you can't think about that every day.*

I read an article in a national magazine about wives at Edwards. It read, "Every day when her husband walks out the door, she wonders if he will come home at night." I just shook my head because we didn't live like that. We couldn't. However, I had a boy in my third grade Sunday school class at the base Chapel whose father ejected from an airplane and was badly injured. The other person in the aircraft was killed.

---•---

AMY ROSS: *I really liked it there. As a kid at Edwards, I felt isolated and protected. It gave me freedom. My brother, Scott, and I would go out and play in the desert. Mom would make us a lunch, and we'd be gone all day long. We heard stories of gold mines in the distant hills and dug in the sand for treasure. There were scorpions and snakes and all kinds of things out there, but we weren't afraid of them. Like the snakes, sonic booms were part of the landscape, and I learned to love that sound.*

After we were there a few months, I started preschool, and we lived at Edwards until I was halfway through second grade. The school was right in the middle of the base housing area. Most of the kids paid a lot of attention to what rank their father was. I didn't even know what rank my father was. I didn't even know what kind of airplanes he flew in. We heard the sonic booms more than we got to see the planes, but one time at an air show I got to sit in some of the cockpits. I was instructed by Dad in no uncertain terms, "Sit there and don't touch anything." My parents got a photo. Then I was summarily removed.

One of the really neat events at Edwards was seeing the prototype Shuttle Enterprise in 1977. Enterprise arrived by traveling overland all the way from Palmdale, California, on a specially built flatbed transporter pulled by very large trucks. NASA took Enterprise up for drop tests five times with astronauts on board to land the Shuttle on the dry lake bed and concrete runways. The whole family went out to watch it land. We stood on the dry lake bed as the Shuttle came in, and Dad was so excited that he was running along the lake bed taking photos. I remember Mom yelling to him, because he was about to fall into a creosote bush as he ran along, shooting photos, not looking where he was going. Mom ran alongside him saying, "Don't trip and hurt yourself; you're going to hurt yourself!" It was exciting, mainly because Dad was so excited.

At that time I didn't know what an engineer was, but I knew Dad worked really hard. While he was in school, he was always studying, and we had to keep quiet. The first Christmas at Edwards, Dad had worked so hard that he got sick. He crawled out of bed, sat propped up against the wall long enough to watch us open presents, and then crawled back to bed.

Dad bought Estes rockets for Scott and me. He helped us build them. Then he took us out into the desert to launch our rockets. But before we launched, we had to make sure there weren't going to be any low-flying aircraft in the area. It was glorious to watch the rockets leap into the bright blue desert sky. I only launched rockets—no animals like Dad had. I'm too much of an animal lover.

In addition to watching the Enterprise landing tests, one of the experiences I really remember about those years is family stargazing. The sky out there was so clear, and the stars were really bright. We'd go out away from inhabited *areas of Edwards where there were no lights, and we'd all lie on blankets, look up at the sky, find constellations, and look for shooting stars.*

———————————————●———————————————

Watching the magnificent starry nights and meteor showers out in the desert of California reminded me of when Jim Gentlemen and I lay on stacks of hay bales and looked up at the sky over Indiana. A lot had happened in my life since then, but I was still looking up at the stars, hoping to go into space.

Flying all the time in Test Pilot School, I quickly got over airsickness. I can still get sick in an airplane if we do a lot of acrobatics, but fortunately, my airsickness does not translate to space sickness. I never had a problem going up or coming down in a Shuttle. There seems to be no correlation between motion sickness in airplanes and space sickness, and no one can predict who is going to feel sick in space.

In Test Pilot School, I primarily flew in T-38s, T-33s, RF-4Cs, C-130s, and KC-135s. Altogether, I flew in fifteen to twenty different types of aircraft. Sometimes I flew with classmate and future astronaut Brewster Shaw. I would later fly with Brewster in space. Mike Mullane, an Air Force navigator, was a fellow student flight test engineer in my class, and we would fly in space together, too.

The Test Pilot School academics and flying were both very challenging. It was the equivalent of another master's degree completed in ten and a half months, and I enjoyed it, especially the flying. In July of 1976, I graduated as the Outstanding Flight Test Engineer in my class, which allowed me some freedom to shop around and see what assignment I would like to have next.

The B-1A Lancer bomber, the Air Force's highest priority program at that time, was being tested at Edwards. Flight test engineers routinely fly on test flights of test aircraft, but they are not normally required. The B-1A was one of the very few airplanes ever that required a flight test engineer to be on board for every test flight. That's where I wanted to work.

The B-1A was an incredible plane. It was designed for supersonic, high-altitude flying and also for hands-off, low-level, terrain-following flying. I liked the fact that we were doing cutting-edge flying.

We had three test aircraft when I worked on the B-1A program. Aircraft 1 and 2 had a crew compliment of pilot, copilot, and flight test engineer. Aircraft 3 had an additional seat for an offensive weapons operator/navigator. The pilot and copilot sat up front, with the other seat or seats separated from them by about six feet. Back where my seat was, behind the pilot, there weren't any windows, so I frequently leaned out into the aisle to look forward and out the front windows. When my duties allowed, I would get out of my seat and stand immediately behind the pilots. In Aircraft 3, there was equipment above that made the aisle more like a tunnel, and I couldn't see the pilots or out the windows from my seat.

The test team conducted a briefing the day before each flight, and we showed up early the next morning to fly. We climbed the ladder into the airplane, started the four General Electric F-101 engines, performed functional checks of the aircraft's flight controls, and taxied to the runway. The moveable wings were set to their full forward position of fifteen degrees, and the slats on the leading edges of the wings, along with the flaps on the back of the wings, were set for takeoff. When the afterburners on the engines were ignited, even at the heaviest takeoff weights of nearly 400,000 pounds, the B-1A accelerated quickly and climbed into the usually cloudless Edwards sky.

One of my main onboard functions was to operate the tape recorders that captured all the information generated by extensive instrumentation on the aircraft. I also monitored the weight of the plane, its center of gravity, the angle at which the moveable wings were set, the speed and altitude, and many other requirements or constraints. The flights normally lasted six hours. Often I would recommend to the pilots and the ground control team that the sequence of the planned testing be shuffled or some backup test points be inserted because the aircraft's gross weight or other factors I was tracking made that change advantageous.

We did most of our flight testing over the Pacific Ocean. We conducted tests to verify the structural integrity of the plane, to assess its flight control systems and its stability, to verify that it did not have any unsafe flutter characteristics, and much more. We tested the engines throughout all combinations of altitude and airspeed. We tested the aircraft's terrain-following radar and the offensive avionics systems.

The airplane was comfortable to fly in, and the large bomber's performance was akin to a much smaller fighter-type airplane. I liked the supersonic testing and the low-level, terrain-following tests the most.

I was on the crew for the only flight on which tests were performed at the aircraft's maximum dynamic pressure. The maximum dynamic pressure is a combination of the aircraft's maximum speed and the lowest altitude at which it can fly at that speed. The testing was conducted over the Pacific Ocean. We refueled from an Air Force tanker as the B-1A flew north along the California coast. The refueling was completed near the border of California and Oregon, and the aircraft was turned southbound. The pilots pushed all four throttles to maximum thrust. We accelerated and climbed. As the plane exceeded Mach 1, the speed of sound, we started to hear abnormal whistling noises inside the cockpit. We continued to climb and to accelerate until we reached the desired test conditions of Mach 1.92 and approximately 36,500 feet. We were traveling at about 19 miles per minute. The whistling sounds became even more unusual, as if the airplane was trying to tell us that it was really working hard at this speed and altitude. In a very short time we completed the planned series of tests at those conditions. We decelerated and turned back northbound to avoid flying into and disrupting the very crowded Los Angeles airspace.

The low-level flying was extremely interesting. It entailed flying at only 200 to 500 feet above the ground and traveling at 0.85 to 0.95 Mach, over 650 miles per hour. To put that into perspective, we flew these big planes at nearly the speed of sound at a height of about the length of a football field off the ground. That would get the attention of the crowd in a flyover before a game!

The B-1A was designed to fly at these speeds and altitudes day or night without the pilot's hands touching the controls. The B-1A used a forward-looking radar that told the flight control computers where the ground was ahead. Downward-looking radar altimeters told the computers how close we were to the ground directly below. The aircraft's computers also had the route that we wanted to fly over the ground programmed into them. On some tests we flew over relatively flat ground and directly at a single hill or mountain peak. We needed to see how the radar and the computers performed in detecting the sudden rise in the terrain and then properly commanding the aircraft to fly up and over and back down once past the peak.

The most dramatic tests of all were the ones we flew over and through the Sierra Nevada Mountains. We would find ourselves flying through valleys in the high country with rugged granite walls on both sides of the aircraft. The B-1A flew exactly as it had been designed to fly. Those tests were the most beautiful and the most exciting. Sometimes when we flew through the canyons, I wondered if the wingtips were going to come through with us.

I had twenty-three test flights on the B-1A bomber, and several of them were terminated early due to emergencies. Parts would fall off the airplanes and equipment would break. We were pushing the limits to figure out what that aircraft could and could not do.

When I was not on board, I frequently served as the test conductor in the ground control room. The test conductor's job was a mix of the duties that the Flight Director and the Capsule Communicator perform in NASA's Mission Control. The control room was filled with strip charts and monitors where engineers stood analyzing every aspect of the test data being relayed to the ground. As test conductor, I talked to the crew and coordinated with all of the engineers in the room.

Frequently, the tests would be performed in a "build-up" fashion. The pilot would make a small test input on the B-1A, the engineers would analyze the airplane's response in real time, and the ground team would approve a larger input. I would relay the information to the flight crew, and the process was repeated until the desired maximum test condition was satisfactorily completed. This was fun, challenging work and was—almost—as good as flying.

When I was test conductor, I also coordinated the refueling tankers and the chase aircraft. The chase aircraft followed the B-1A to visually observe the testing and provide support if something happened, like a piece coming off the airplane.

As the lead Air Force engineer for evaluating the B-1A's flying stability and the effectiveness of its flight controls, I spent many hours crunching data on computers and analyzing the results. I also helped plan upcoming test flights. There were always more tests that engineers wanted to perform than could ever be accomplished. This necessitated a continual review of what tests would have the highest priority for the next flights. All of this made laying out an efficient test plan with proper aircraft configurations and flight conditions for each test point, with as little time as possible between tests, an intricate job.

One year after I started working on the B-1A, NASA announced that it would hire the first class of astronauts for the Space Shuttle program. It was 1977, and the last astronaut group had joined NASA in 1969.

The mid-to-late 1970s was a quiet time in US spaceflight. The last Apollo mission—Gene's Cernan's Apollo 17—had safely landed in the Pacific Ocean in December of 1972. Since then there had only been three manned missions to NASA's Skylab in 1973 and 1974, and one joint Apollo-Soyuz mission with the Soviets in 1975.

Just about everyone I knew at Edwards scrambled to get their applications in, as did nearly all the test pilots and flight test engineers at other flight test centers around the country. Ultimately, more than 8,000 people applied, including Jerry L. Ross.

I had to apply through the Air Force, and they selected my application as one of those to be forwarded to NASA. I was very excited when I received a call from NASA to be one of 210 applicants invited to Houston for interviews and physicals. I was getting close.

NASA brought the interviewees to Houston in groups of about twenty. They put us through a pretty normal set of medical tests—easy compared to what the first Mercury astronauts went through. The flight surgeons checked our eyes, ears, noses, throats, hearts, and lungs. They explored parts of my body I didn't even know I had, and then they put us through an exhausting treadmill endurance test. NASA also checked our past medical records extensively, though most of us had already been through many medical screenings to do our current work. We were in good shape.

There were two psychiatrists who separately talked to each of us, using the good cop, bad cop technique. One psychiatrist wanted to see if we would lose our composure under rapid questions and psychological testing. The other psychiatrist was laid back and acted like he was our grandpa. I think his hope was that we would reveal secrets from deep in our souls. I was careful with both of them.

During the week, NASA also put each interviewee into a prototype rescue sphere. The engineers explained that they were considering it for use on the Shuttle, and they wanted each of us to evaluate it and write a brief report. The actual purpose of the test was to determine if we were claustrophobic. We had to double up to get into the sphere. Then they zipped it closed and did not tell us how long we had to stay in there. There was no window in the sphere, and air came in through a hose.

I fell asleep.

We all got a tour of the Johnson Space Center and were given several informative briefings. The most important event for each of us, however, was an individual one-hour interview with about a dozen people around a conference table. This was the Astronaut Selection Board, composed of astronauts and NASA managers. Board members asked questions starting with our childhood, and the interview proceeded from there. They asked a lot of questions. They would interrupt our responses to see if we could pick up our thoughts again. The interview with the Board was primarily a way for NASA to find out if the interviewee was the kind of person they wanted to have working and flying with them. NASA is very particular about who they put in their spacecraft.

I walked out of the interview thinking I had done a great job!

The evaluation of candidates went on for months, so there was a long, and on my part impatient, wait. At Edwards, we heard through the grapevine that if you had been selected, the head of the Astronaut Selection Board, Mr. George Abbey, would make the phone call. If you had not been selected, one of the other Board members would call.

I finally got my phone call. It was from Ed Gibson, who had flown on the final Skylab mission. I don't remember what he said, but it hurt.

Setback number three, and it was the worst one of all. I was so close. I thought I had achieved my goal, but then it was snatched away.

Six flyers at Edwards got calls from Mr. Abbey. Test Pilot School classmate Brewster Shaw was selected. Ellison Onizuka was also on the list. He was an instructor at Test Pilot School, and we played on the same softball team. El and I had gone to Houston together for our interview week. Another softball teammate, and a Purdue graduate, Loren Shriver got the call. The other three were Air Force test pilots Dick Scobee and Steve Nagel and Army test pilot Bob Stewart. The new Class of 1978 was a huge one, and in Houston they were nicknamed the TFNGs—Thirty-Five New Guys. NASA's Group 8 included the first female, first Asian American, and first African American astronauts. Eight years later, four members of this class, including Dick and El, would be among the seven who died on the Space Shuttle *Challenger*.

When I saw my friends leave for Houston, it was hard.

I did a lot of thinking and praying about what to do after not making the cut at NASA. I talked with Karen and with some of my friends who had been selected. And I commiserated with others at Edwards who, like me, had missed the cut and were standing on the outside looking in.

Everyone I worked with knew I had interviewed for the astronaut program, and they knew I was disappointed. One of them, Doug Benefield, had already been in my shoes. Doug was a highly experienced test pilot who was working on the B-1A for Rockwell. He and I had hit it off from the beginning. I really enjoyed his company and flying with him, and he seemed to enjoy having me around. He had applied to be an astronaut during the Gemini program and didn't make it, so he knew what I was going through.

That summer Doug invited Karen and me to join him for the Society of Experimental Test Pilots banquet in Los Angeles at the end of its annual symposium. Karen and I were excited to be in the same room with so many highly experienced and respected aviators. The highlight of our night was meeting retired World War II Lieutenant

General Jimmy Doolittle, who commanded the first strike against the Japanese homeland after the bombing of Pearl Harbor. Jimmy Doolittle was a man of small stature, friendly, and he had a wonderful smile. He lit up the room. Meeting this living piece of history was a great honor.

Doug was a great friend and a terrific pilot. I was devastated when he died on August 29, 1984, in a B-1A bomber crash during a test flight at Edwards. I still miss him and his impish smile.

I wanted to know if there was any chance I could still be an astronaut, so I decided to go to the top. During the next three months, I worked up the courage to call Mr. Abbey, which I finally did. I started by saying, "I'm trying to figure out what to do with the rest of my life. I would like to know if you saw something that would preclude me from getting another interview opportunity."

Mr. Abbey told me the Board hadn't seen anything that would prevent me from being considered in the future. In fact, he said he hoped I would apply again. I was excited to hear his encouraging words.

I asked him for advice, laying out my options. I told him that I was in line to become head of the Test Pilot School's Flight Test Engineering curriculum. The Air Force also had offered me an opportunity to get a PhD and become an instructor at the US Air Force Academy or to go straight to the Academy to teach. A friend that I had worked with at Edwards had been placed in charge of one of those aircraft test programs that "don't exist." He wanted me to join him in top secret work with the enticing promise of "long hours, low pay, and lots of flying." That sounded really good to me. But it wasn't space.

"Those are all good options," Mr. Abbey said. "But I'd like for you to consider one more. Why don't you come to Houston as a military detailee and help us integrate military payloads into the Space Shuttle?"

Then there was a pause.

"No promises should be expected and none are being made," he continued. "But it would give us a better chance to get to know you, and it would give you a better understanding of the NASA organization and what the astronaut job is all about."

I knew I was going to Houston before I hung up the phone, though it was going to be difficult to give up the great job of flying on the B-1A.

In fact, if I left Edwards, I didn't know if I would ever fly again in high-performance military aircraft, but I had to continue to follow my dream.

It took almost a year to get the official paperwork from the Air Force assigning me to the Johnson Space Center. Once I got the orders, I began making arrangements to move.

Amy and Scott told me what they wanted in our house in Houston. They each wanted their own bedroom. The house had to be two stories, and it had to have a fireplace. I had my marching orders, and I went on a four-day house hunting trip.

The real estate agent showed me a number of places, most of them close to the Johnson Space Center where many NASA people lived; however, I was looking for something a little further out, more rural, and I knew Karen would like that, too. Finally, I found a house in Friendswood that looked right. It met all the kids' requirements, so I bought it. Karen and I still live in the same home, but the surrounding area is now filled with houses, and it doesn't feel like the country anymore.

While I was excited about going to Houston, I was sorry to leave the wonderful friends we had made. I enjoyed the flying at Edwards, and I have tremendous respect for the people I worked with there.

During my farewell party, coworkers put on a skit. The guy playing me would frequently walk in, interrupting the ongoing scene, and say, "Did Houston call? No? I'll see you later, I'm going flying."

I guess they knew me pretty well.

Our family moved to Texas in February of 1979. Since we already had the house and the Air Force-contracted movers couldn't arrange for our belongings to arrive at the same time we did, we moved ourselves. I loaded everything we owned into a U-Haul truck, and we headed toward Houston.

AMY: *Dad and Scott were in the U-Haul truck, and Mom, our dog, Prancer, and I followed in the car. When we got just east of El Paso, the truck broke down. The radiator needed water. There was water in the back of the moving* *van, but it was buried underneath everything. Being the smallest and most flexible, I was selected to crawl around everything and find the water. The rest of the trip was a little nerve-wracking, keeping our fingers crossed that the truck would make it. When we got to the Houston area, I wasn't sure what to think. We drove past chemical plants and oil fields, and there were rice paddies across the street from our house! However, the house was two stories, it had a fireplace, and I got my own room. I also got orders from Scott that no girls were allowed in his.*

Our first Sunday in Friendswood we attended services at Friendswood United Methodist Church. We entered by a side door and met a bald man making coffee in a small kitchen area. He introduced himself as John Barfield and showed us to the sanctuary. A little later as the service began, that bald man entered the sanctuary in a robe and went up to take his seat in the pastor's chair. After a few months I asked John to put me to work. I served as the stewardship chairman for the next two years. Since then, Karen and I have served in many positions in the church, and Karen has taught our Christians in Action adult Sunday school class for over fifteen years. We have participated in mission trips to Homestead, Florida; Costa Rica, Mexico, and Kenya. We have been blessed by our church family and have enjoyed our opportunities to serve.

The week after we moved in, the kids brought a note home from school that said everyone was to wear Western clothes to school that Friday. Scott had jeans and a long-sleeved plaid shirt, and luckily Amy had a little red outfit with white braiding that she could wear. Amy and Scott had already told us they were singing "The Eyes of Texas" and "The Yellow Rose of Texas" at school.

Like everyone in the country, we knew about the stereotype of Texans being boastful and proud of their state. We thought that was an exaggeration, a caricature, but our children were being immersed in it at school. It seemed Texans took Texas pride very seriously. Only after being in Texas a month did we learn that particular Friday was Go Texan Day, the day before the Downtown Rodeo Parade and the kick-off of the annual Houston Livestock Show and Rodeo, the day when people in the Houston area pull out their Western duds. We had arrived in Friendswood at just the time when Houston wears its cowboy hat and boots and celebrates its Texas heritage every year.

We used the money earned by performing the move ourselves to buy a refrigerator. Since that time more than thirty-three years have passed. I've completed seven Shuttle flights and nine walks in space. Our children have grown and gone.

And that refrigerator is still working in our kitchen.

KAREN: *That spring I worked as a substitute school cafeteria worker, a part-time clerk at a fabric store, and a substitute teacher.*

I was hired to teach Texas history and life science at Friendswood Junior High School the next school year. Because we were in a different region of the country, I had to take additional courses to get a Texas teaching certificate. I also wanted to take a course to learn some Texas history before I taught it. I started taking classes in May. Several nights a week, as soon as Jerry got home to be with Amy and Scott, I went to class. We were both very busy.

When I started work, I learned Mr. Abbey had also hired several other people who had interviewed in 1977 and hadn't been selected. The others arrived at NASA earlier and had laid claim to every good job. I was the new kid on the block, and my managers kept handing me stacks of things to read. I started to wonder if I had made the right decision, giving up flying for all this reading.

El Onizuka and Loren Shriver asked me to join their softball team at the Johnson Space Center. I showed up for the first practice and discovered "their" team was actually the Astronaut Office team. Mr. Abbey was the catcher. I played in the outfield and got to know more of the TFNGs.

After several months I was named the payload officer and flight controller responsible for the operational integration of all military payloads being planned for future Shuttle missions, and that kept me busy. In my role as payload officer, I worked with many groups around the Johnson Space Center and got to know some of the people who had been part of the Apollo program. I met Gene Kranz, the legendary Flight Director who led Mission Control's efforts to land Apollo 11 on the Moon and to safely return the crew of Apollo 13 to Earth. Gene was a former US Marine fighter pilot, and his flat-top haircut, broad shoulders, and powerful personality embodied the tough and competent approach that he instilled in his Flight Directors and his flight control teams.

Gene's legacy still guides the activities in the Mission Control Center supporting the International Space Station operations twenty-four hours a day, seven days a week. His book *Failure Is Not an Option* is an excellent read and really captures the challenges and excitement of America's early manned space programs.

In 1979, many people, like Gene, who had worked on the Apollo and even Gemini programs were still at the Johnson Space Center, but their numbers were dwindling. The Shuttle was behind schedule, and a US-manned mission had not launched in four years. Some people compare this time period to the current post-Shuttle era, but there's one big difference. In 1979, the country knew exactly what we were going to do next in human spaceflight. Today, it's very uncertain.

In 1979, NASA was also dealing with Skylab's looming and unwelcome return to Earth. Skylab had been launched into space in May of 1973 atop a Saturn V rocket. It was the last time that huge rocket that sent Americans on their voyages to the Moon was launched. After the three manned missions to Skylab ended, its orbit slowly deteriorated. In 1977, NASA considered sending a Space Shuttle to Skylab to lift it to a higher orbit and extend its useful life. Program managers thought this mission could be accomplished in 1979.

However, the Shuttle wasn't ready in 1979, and it became only a matter of time before Skylab would reenter the atmosphere and fall to Earth. Exactly where it was going to hit became an international concern, with some people looking warily at the sky while late night TV comedians cracked jokes about it. On July 11, 1979, the remains of Skylab crashed into the Indian Ocean and onto a farm near Perth, Australia. Karen and I stood in our front yard that morning shortly after sunrise and watched the extremely bright and fast-moving Skylab zoom by on what I believe was its next to last orbit of the Earth.

———————————————●———————————————

KAREN: *By the next school year I was teaching seventh graders and going to night classes. Amy and Scott were going to school, doing homework, and playing with neighborhood friends. Jerry was working, playing softball, and preparing his second application to be an astronaut.*

———————————————●———————————————

In 1979, NASA announced another astronaut selection, and I anxiously applied for the second time. My three coworkers in the Payload Operations Office who were interviewed in 1977 also submitted their applications. We all knew the probability that even one of us would be selected was small. This time NASA received 6,000 applications and interviewed 120. The Shuttle still had not flown.

I made it to the interviews once again. I passed all of the medical tests, but I did not feel the interview with the Astronaut Selection Board had gone well. I walked out of that conference room and thought, "Ross, you idiot. You just blew it. You might as well pack up your bags." All those years of work were going down the drain. And I had given up the B-1A for nothing!

The phone call came on May 20, 1980. I was in Mission Control when they tracked me down.

This time the call was from Mr. Abbey.

I don't remember much of what he said, but I do remember jumping up and down. Everybody looked at me. I was smiling from ear to ear, and then they all figured it out and congratulated me.

At the Johnson Space Center there's a large, green, open park area with ponds between Building 30, where Mission Control is located, and Building 4, where my office was situated. I floated across that grassy mall. Every time I walked that path afterward, I recalled how I felt after that telephone call.

I returned to the Payload Operations Office and found my coworkers Mike Lounge and Bonnie Dunbar had also received good news. We were amazed that three of the four of us had been selected. I was also pleased to see three of the others selected, John Blaha, Roy Bridges, and Guy Gardner, were fellow Purdue graduates. Another person on the list was Charles Bolden, who was the NASA Administrator when I retired on January 20, 2012.

I called Karen at work, and she was overjoyed and relieved. Amy and Scott were visiting in Indiana, and I called them at Karen's parents' house. After I talked to Amy, she hopped up on a barnyard fence and walked back and forth along the top rail, repeating over and over, "My daddy is an astronaut, my daddy is an astronaut . . ."

I called my folks, and they were excited though apprehensive. I think it just began sinking in for them that I might actually be launched into space one day.

Being selected on the second try was not bad. There are people who have applied seven times before making it—or not.

I was thirty-two years old, and I had accomplished my life's goal—sort of. I still hadn't actually flown in space, but all the years of working and saving money to get to Purdue, all the years of study, all the hard work during my Air Force career had paid off. I felt incredibly happy. And thankful. Thankful to my wife and family. Thankful to God.

The road to this point had not been straight. There was the Air Force ophthalmologist who kept me from becoming a jet pilot. Karen's car accident had kept me at Purdue, which resulted in me getting a master's degree in mechanical engineering before I entered active duty in the Air Force. There was the declined assignment to Test Pilot School. And when I thought my big opportunity to become an astronaut had finally arrived, I didn't make it.

I have learned that you are not going to get to do all the things you want to do in life when you want to do them, if at all. Life is full of trials and twists and turns, and you have to play with the cards you're dealt. You have to rely on your support system—your family and your faith. God has a plan, and I believe you can subvert God's plan by overreacting to disappointment, often without realizing that disappointment actually may be part of His plan.

I had also learned that when things don't go right, you just have to go on. I felt blessed to have had so few setbacks, and I chose to see them as just being delays.

More than that, what would have happened if things had always worked out the way I wanted? If I had gone to Test Pilot School the first time it was offered, events might have unfolded very differently. I might not have graduated first in my class. I might not have gotten the assignment with the B-1A bomber program. If things had gone any other way, I may not have reached my desired destination.

And there was something else that I wouldn't learn about until years later. As it turned out, one of the colonels managing the B-1A bomber test program at Edwards had been a classmate of Mr. Abbey at the US Naval Academy.

That officer was Colonel Dick Smith. Knowing Mr. Abbey, I feel fairly certain that sometime before I was selected, Mr. Abbey called Colonel Smith to ask about my performance on the B-1A program. Evidently, if Dick Smith was called, he said some nice things.

Shortly after I was selected as an astronaut, Reverend John Barfield asked me to give a talk to our church youth group. John thought my new position in life would give me great credibility with the teenagers. I struggled with what to say and prayed for help. After a week or so, the outline of a presentation began to take form in my head. I started putting my thoughts on paper.

Soon I had put together the story of my life, how I had worked in concert with God's plan for me, and how God had helped me to become an astronaut. Since then I have given speeches in all fifty states, Washington, DC, and in seventeen foreign countries. That earliest version has changed little over the years and has been incorporated into most of the talks that I have presented to groups of children and adults all over the world.

This is the core of the message. I believe we are all here for a purpose. God makes each of us a unique, special person with a unique set of likes, dislikes, talents, and gifts, and it's up to us to figure out what they are. Then we can select a career path best suited for us, a career path that enables us to use those talents not just for ourselves, but for others as well.

I also believe we can't leave it all up to God. You have to work hard yourself. I believe in setting goals, studying hard, and working diligently to achieve your goals. I believe we should not give up easily. I believe we have to keep focused if we want to succeed. If you use your skills in the manner God intends, you will be good at what you do, and you won't feel like your job is work. You will enjoy your work because it's something you were designed to do. You will succeed because you are working in conjunction with God's plan.

God's hand has guided me and helped me achieve my goal of being an astronaut. He will guide you, too.

Becoming an astronaut was a turning point in my life. Everything I had done up to that moment was focused on that singular objective. And everything I would do from that moment on was focused on becoming the best astronaut I could be.

Now there was only one problem.

Astronauts have to swim.

FOUR

The no-names

On my first Shuttle flight, STS-61B, we launched at seven thirty at night under a bright, full Moon and into a clear, starry sky. Out the front windows as we climbed toward orbit I could see the constellation Orion, "the hunter," with its three-star belt.

It was the same Orion I had seen in the northern Indiana sky as a boy getting up before dawn on winter mornings to check traps in the marsh; the same Orion I had pointed out to Karen, Amy, and Scott as we lay on blankets during crystal clear nights in the California High Desert at Edwards Air Force Base.

But I had never seen it like this.

When I gazed at Orion before, I was just dreaming of going into space. Now I was most definitely on my way, heading from sitting on the launch pad to a speed of 17,500 miles per hour in eight and a half minutes—that is all the time it took to get into space in the Shuttle.

It seemed like both the shortest and longest eight and a half minutes in my life.

Even though we launched at night, it was light inside the cockpit all the way to orbit because of the fire from the rocket engines. At one point during the ascent, I looked backward through the overhead windows and could see flames blasting from the Shuttle *Atlantis'* main engines. The flames were not steady and conical as they are in artists' renditions. They looked more like candle flames flickering in the wind, almost disappearing at times. The flames were dancing.

I faced forward again and looked at the instruments to make sure everything was working correctly. I was thinking, "Are the engines okay?"

People always want to know what the Earth looks like from space. How many ways can you say "beautiful"? I distinctly remember my initial peek at the ground as we passed over Africa on my first orbit. I could see many fires, and I guessed the people down below were burning off fields in preparation for a new crop.

One of my favorite things to do in space was to look down at God's creation. From our normal orbital altitude of about two hundred miles, I could see more than one thousand miles in any direction. I could see oceans, continents, coastlines, lakes, rivers, mountain ranges, volcanoes, and weather systems. The Earth passed by as a continuously changing mosaic of colors, textures, and shapes. I had studied world atlases, and looking down, I had a good understanding of where I was and what I was seeing. It was incredible to silently float over the coast of Florida and see the beautiful blues and greens of the Caribbean, followed just minutes later by the coastlines of Africa and the Mediterranean Sea, the Nile River and Delta, the Dead Sea, the Arabian Sea, the Himalayas, the purplish tint of Australia, and the wide expanse of the Pacific Ocean before crossing over the Baja Peninsula of Mexico and over the United States again.

The southern island of New Zealand is among the most spectacular sights from space. It's absolutely incredible—stark snow-covered mountains dropping steeply to the deep blue waters of the South Pacific. It made me want to visit what has to be one of the most beautiful places on Earth.

From low Earth orbit one can see a sunrise and a sunset every ninety minutes, each more vivid, unusual, and beautiful than the last. Within a few minutes each sunrise rapidly transitions from complete darkness to exceedingly brilliant daylight. One of the most unexpected parts of sunrises and sunsets is the layering of the atmosphere above the Earth's horizon, much like the layering of a torte. The layers have distinctively different hues of purples or blues or oranges or reds, and they are constantly changing. I have counted as many as thirteen layers of color.

To see stars at night, we had to turn down all the orbiter's interior lights to make it dark and eliminate the reflections in the windows. Then we could view so many more stars than one could ever gaze at from Earth, and during a single orbit we could see both Northern and Southern Hemisphere stars.

You might think that looking at the Earth from space at night would be a waste of time. But at night, lights on Earth are impressive and make identifying cities and major roads easier. Lights from land and the absence of lights on water distinctly define coastlines and the large population centers located along them.

One of the most awe-striking views is a thunderstorm at night. On the ground, thunderstorms can be very dramatic. But from orbit, watching thunderstorms is an entirely different experience. You don't hear the thunder, and you aren't in the path of the coming wind and rain. What you do see are flashes and long streaks of lightning. The lightning illuminates the clouds and reveals the large scale of the storms, which can cover hundreds if not thousands of square miles. Sometimes when watching a thunderstorm, it appears to almost stop. Then one lightning bolt strikes, and that bolt seems to trigger a chain reaction of lightning that erupts along hundreds of miles of the storm front. I saw this phenomenon for the first time in a large storm that covered much of South America. The interplay of lightning over such large distances was a great surprise. This display of power totally changed my understanding of and appreciation for these forces of nature.

That first launch into space was an incredible moment, the fulfillment of a lifetime dream. But it didn't come quickly after being selected as an astronaut by NASA. It was more than five years from the time I was selected until I participated in my first flight.

As astronaut candidates, called ASCANs, we knew from the start that the wait would be long. NASA hired our class in 1980 before the first Shuttle flew. There were twenty-five Apollo-era astronauts lined up for missions, and there was that whole class of thirty-five astronauts selected two years before us waiting for their first flights. There was also a lot to learn about being an astronaut before we were ready for space.

One of the senior astronauts in the Astronaut Office at that time was Al Bean, who was on Apollo 12 with Pete Conrad and was the fourth man to walk on the Moon. Later, Al commanded the Skylab 3 mission.

Al gave me one of the most important pieces of advice that I ever received about being an astronaut. He said, "When you're out doing public appearances, never pass up the opportunity to use a restroom. You don't know when you'll see one again—or have the time to use it." I've never forgotten his guidance; frankly, it has saved me on numerous occasions.

Al was the first and one of the few people who talked to our class about all the personal challenges we would face as astronauts, and we always appreciated his sincerity and his willingness to share his lessons learned, some of them the hard way. He also talked to our spouses.

———————————————— ● ————————————————

KAREN ROSS: *During our orientation, the spouses of the astronaut class of 1980 were given a nice tour of the Johnson Space Center, and Al Bean gave us a briefing. Al had asked NASA managers for an opportunity to talk to us.*

He told us about the intense pressure the astronauts are under before a mission. Al wanted us to understand that as a crewmember, every astronaut is keenly aware that the decisions he or she makes can not only jeopardize the success of the mission, but also cost everyone on the crew their lives.

Al spoke to us very specifically about what we could expect. He said, "I want you to know that when the time comes for your husband to fly, he may not act like the person you know. Maybe he has always taken care of the car. If the car breaks down, he may ask you to take care of it. Maybe your husband has always repaired things at home. But if you have a problem with the washing machine when he is preparing to fly, he may tell you to buy a new one. For that time he may seem detached. All his focus must be on the flight ahead."

Then he related an experience he had while he was getting ready for Apollo 12. He went to a social event, and during the evening, he talked to many people. When the evening was over, he realized he couldn't recall anyone he had talked with. Even though he had carried on conversations, he was thinking all the time about the things he needed to remember during the upcoming flight. But, Al concluded, there is an end to it. He wanted us to know the mission will come and go, and our husband will come home again.

Al shared that he and his wife had divorced. He thought that if someone had been able to tell the wives of the earlier astronauts what to expect, maybe their marriages would have fared better.

Talking so personally and so openly with us was a very caring thing for Al Bean to do. It really made an impression.

Most importantly, he was right, right about it all.

And by the way, years later when our washing machine developed a leak, I wasn't too surprised when Jerry told me we were just going to buy a new one.

In addition to being a Naval aviator and an astronaut, Al Bean is a talented artist. He primarily paints Apollo lunar scenes. Late in my NASA career, I took my whole family, including my three granddaughters, to visit him in his studio in Houston. He showed all of us around and explained the techniques and some of the tools he uses in his paintings. He uses the geology hammer and circular core tube bit that he brought back from the Moon to texture his acrylic paints. He also incorporates small pieces of the patches that were on the space suit he wore when he walked on the Moon. Those patches have lunar dust embedded in them. The paintings are beautiful. It was all extremely interesting, and Al was very gracious. The visit was a special treat from one of the twelve Moon walkers.

Our ASCAN training lasted one year. I enjoyed my classmates and the whole training process. NASA instructors taught us about orbital mechanics, the human body and how it is affected by space travel, space physics, planetary sciences, the scientific principles involved in the types of experiments we might do, and all about the orbiter systems and subsystems. We received training on Shuttle operations. We learned water and land survival skills.

The trainees and trainers alike were in the early stages of the Shuttle program, and we were all learning together. This was the period when NASA was drawing closer to flying the Shuttle for the first time, and the attention of everyone was focused on that event. It certainly was not very focused on the new class of astronauts.

We were the ninth NASA astronaut class. As of 2012, there have been twenty classes, and the twenty-first is scheduled to report to NASA in 2013. All but three of these classes have had nicknames given to them early in their training. Some of these nicknames sound very official, like they might have come directly from NASA management, while others are just plain creative. Astronaut groups have been labeled the Mercury Seven, the Next Nine, the Fourteen, the Scientists, the Original Nineteen, the XS-11 (Excess Eleven), the TFNGs (Thirty-Five New Guys), the Maggots, the Gaffers (George Abbey's Final Fifteen), the Hairballs, the Hogs, the Flying Escargot, the Sardines, the Penguins, the Bugs, the Peacocks, and the Chumps. Many of the names were inside jokes. For example, the Peacocks were teasingly told they had been selected "just for show and not to fly."

My 1980 class didn't get a nickname. Only one class before us didn't get a nickname, and they were a group of military astronauts who transferred into NASA when their military Manned Orbiting Laboratory project was canceled. Not a good omen. In fact, my no-name class was lost in the hubbub to get the first Shuttle flight off the ground.

We signed in at Johnson Space Center the week after July 4, 1980, and the Shuttle program had already suffered multiple delays. When NASA first proposed the Space Shuttle to Congress and the administration, officials described a program with one hundred flights per year. That meant averaging one launch about every three to four days. Frankly, the idea of launching one hundred Shuttles per year was preposterous. It was not feasible then, and it was not feasible in 2011 when the program ended.

By the time we arrived, NASA had cut the projected goal back to fifty per year, which was still totally unrealistic. In those early planning days, NASA severely underestimated the turnaround time that would be required to get each vehicle ready between flights and didn't understand the difficulties they would have getting the Shuttles launched once on the pad.

The Shuttle had been designed with multiple levels of redundancy. It was originally planned that with so many backups in place, the orbiter would be able to launch without everything being fully functional in every subsystem. Using that approach, the designers believed the Shuttle could be treated like a commercial airliner and that only periodic, comprehensive maintenance inspections and tests would be required. The vehicle would land, the experiments and payloads that had returned from space would be taken out, and new ones installed. The tanks would be filled, the tires kicked, and the vehicle would be ready for another launch. There would be no time-consuming and costly disassemblies, inspections, tests, and subsystem-by-subsystem rechecks performed between flights.

In reality, the Shuttle was not a commercial airliner. It was a revolutionary design and the first reusable space vehicle ever to fly. A more conservative approach was prudent. With valuable national assets and human lives at risk, the operational philosophy required that all subsystems had to be completely checked and every backup had to be fully functional before the Space Shuttle was approved for flight. All of these in-depth inspections and tests between each flight were very time-consuming and expensive.

Shuttle operators at the Kennedy Space Center also found that the routine tasks took longer than estimated. For example, instead of quickly cleaning the bathrooms between flights like they do on a commercial airliner, the Shuttle toilet had to be completely removed, thoroughly cleaned, and reinstalled before each flight. And the change-out of experiments and payloads between flights took much longer than originally expected. In addition to the Shuttle itself, processing of the external tank and the solid rocket boosters sometimes caused delays.

Once all the individual elements were ready, they still had to be moved to the Vehicle Assembly Building and put together—or stacked. Mechanical and electrical connections had to be mated and a complete end-to-end test performed before the entire Space Shuttle stack could be moved to the launch pad. At each step along the way, a bent connector pin, a leaking seal, or a missing specialized tool could add hours, if not days, to the turnaround time between one flight and the next.

On top of that, the Space Shuttle program encountered potentially critical hardware problems that had to be corrected or required more frequent inspections. When significant problems were detected, NASA would bring the entire flight schedule to a halt until a fix was developed and incorporated into each of the Space Shuttles. The most notable instances of this were in the wake of the two Shuttle accidents, but there were other instances of extended launch delays due to issues discovered while inspecting, testing, and preparing the Shuttles for launch.

Operational weather requirements were another reason why the Shuttle didn't launch as frequently as NASA originally intended. For a Shuttle to launch, the weather needed to be acceptable at the Kennedy Space Center as well as at least one transatlantic landing site so the Shuttle could land in Europe or Africa in the event of an emergency soon after liftoff.

Twenty-one Shuttle landings were diverted from Kennedy to Edwards Air Force Base in the California desert because of bad weather in Florida. Each time a Shuttle landed at Edwards instead of Kennedy, it took at least an additional month to get the Shuttle ready for its next mission.

Early on in the Shuttle program, everyone talked about making spaceflight routine and bringing down the costs of operations. We would be the truckers of space. Unfortunately, human spaceflight just doesn't work that way. The vehicles are so complex, and the program must make sure they're truly safe to operate every time they fly. Space travel is never going to be like boarding a commercial airliner for New York, much less like a truck ride down the interstate.

Before we ASCANS could be assigned to a mission, we had to get to know the Shuttle and its components like they were our best friends. I spent many hours in class, at my desk, and in my reclining rocker at home studying thick manuals.

The Shuttle stack sitting on the launch pad consisted of four major components. The Shuttle, also called the orbiter, was attached to the large, orange-brown external tank, and the two powerful, tall, white solid rocket boosters were mounted on the sides of the external tank. The orbiter had about the same overall dimensions as a Boeing 737. The payload bay of the Shuttle was 15 feet in diameter and 60 feet long, and a city bus would easily fit inside. It was designed to carry payloads of up to 65,000 pounds into orbit.

The two large solid rocket boosters were 149 feet long and 12 feet in diameter. Each weighed 192,000 pounds empty and contained 1.1 million pounds of solid rocket propellant at liftoff. At two minutes after liftoff, their fuel was consumed and they were released. They parachuted to the ocean and were recovered and reused. The orange external tank held more than 500,000 gallons of liquid oxygen and liquid hydrogen that were consumed by the orbiter's three main engines. The tank was 27.6 feet in diameter, 154 feet long, and weighed about 60,000 pounds empty. It was covered with a foam insulation to keep the super-cold oxygen and hydrogen from boiling off too fast. After the fuel was spent, the tank separated from the Shuttle and burned up reentering the Earth's atmosphere.

The entire Shuttle stack stood 184 feet tall on the launch pad—about the height of an 18-story building—and when fully fueled it weighed an amazing 4.5 million pounds. At liftoff the Shuttle's three main engines and its two solid rocket boosters provided more than 6.5 million pounds of thrust.

The six orbiters were assembled in Palmdale, California, and these were flown piggyback on top of NASA's highly modified Boeing 747 Shuttle Carrier Aircraft to the Kennedy Space Center in Florida. The external tanks were manufactured at a voluminous plant at Michoud, Louisiana, near New Orleans, and shipped by barge to the Cape. The solid rocket boosters were manufactured near Brigham City, Utah, and shipped in segments via a special train to the launch site in Florida.

Once at the Cape, the Shuttle components were prepared for flight and transferred to the gigantic Vehicle Assembly Building where the stack was put together on a large mobile launch platform. When everything had been fully assembled and thoroughly checked, the stack and the mobile launch platform were carried to the pad by a monstrous device called the crawler transporter. The crawler transporter weighed six million pounds and carried its eleven-million-pound load at a maximum speed of one mile per hour. It took six to eight hours for the Shuttle to reach the launch pad. The Vehicle Assembly Building, the mobile launch platforms, the crawler transporters, and the launch pads had all been used for the Apollo program and modified for the Shuttle.

The Shuttle had three large rocket engines that helped lift it into orbit. In addition, it also had two smaller orbital maneuvering system engines and fifty-two small reaction control rocket engines. The orbital maneuvering system engines provided six thousand pounds of thrust each and gave the Shuttle the final little push it needed to enter orbit. They made it possible for the Shuttle to rendezvous with satellites and space stations, and they provided the thrust to deorbit from space. The reaction control rocket engines were used to control the orientation of the Shuttle on orbit and during the early portions of the reentry and also for performing rendezvous and docking.

The Shuttle's crew compartment had two levels, the flight deck above and the mid-deck below. The commander, pilot, and mission specialists one and two were seated on the flight deck for launch and landing. There were six windows in front of the pilots, and controls, displays, switches, and circuit breakers covered the walls and the ceiling. There were more than 1,000 switches and circuit breakers in the Shuttle. We knew them all.

On the flight deck the commander's seat was front left and the pilot's was front right. Mission specialist one sat behind the pilot, and mission specialist two sat between and behind the commander and pilot. Mission specialist two served as the flight engineer, assisting the commander and pilot with the checklists and monitoring the operation of the Shuttle systems during launch and landing. The mission specialists' seats were removed and stowed once the Shuttle was on orbit.

The back part of the flight deck was called the aft flight deck. Controls for the television cameras, the robotic arm, the communications systems, the payload bay doors, and for launching satellites from the payload bay were located in the aft flight deck. The controls for the docking mechanism that was used to mate with space stations were also located there.

In the aft flight deck there were two windows looking out into the payload bay and two more in the ceiling. The aft windows were used for many purposes, including the operation of the robotic arm. The overhead windows were also used for many activities, but, most importantly, the commander looked out of them to see a space station as he or she maneuvered to and docked with it. All of these windows allowed for great sightseeing if and when the crew had time available to enjoy the views.

The mid-deck was a true multipurpose room. On orbit we cooked, ate, worked, exercised, bathed, and slept in the mid-deck. The hatch used to enter and exit the Shuttle was on the port (left when facing forward) wall.

The crew entered the Shuttle through the hatch into the mid-deck. Immediately on the right was the toilet. Just inboard and above one's head was the opening leading to the flight deck. The space by the hatch and the toilet was the area for sponge baths. Turning to the left and around the corner was the galley. The galley, or kitchen, consisted of a rehydration station to add water to drink and food packages, and an oven with a fan in it to circulate the hot air in zero gravity.

Continuing clockwise, the front wall of the mid-deck was covered with lockers that held our food, clothes, tools, supplies, and experiments. The starboard (right when facing forward) wall was where our sleeping bags were stowed during launch and landing and also where we stowed our orange launch and landing suits while we were on orbit. On the back wall next to the starboard wall were some additional lockers. The hatch to the airlock was between these lockers and the toilet. This was our port to go on spacewalks. Our space suits were stowed in the airlock for launch and landing. There were also controls and equipment mounted on the ceiling and under the mid-deck floor. Seats were mounted on the floor of the mid-deck for launch and landing. Mission specialists and payload specialists sat in these. Once in space, these chairs were removed and stowed to make room for on-orbit activities. A treadmill or a cycle ergometer for exercise also was mounted on the floor.

Three ingenious devices called fuel cells provided the electrical power for the Space Shuttle. Fuel cells had been the power supply for the Gemini and Apollo capsules. Their unique characteristic was the ability to combine oxygen gas and hydrogen gas in a chemical reaction that created electricity and left pure water as the only by-product of the reaction. The water was collected in supply tanks and used to help cool the Shuttle's systems, to drink, to rehydrate our food, and for sponge baths.

The Shuttle was a truly amazing vehicle.

As we neared the end of our year as ASCANs, each of us was asked to choose what type of support work we wanted to do from a list of options. I had always wanted to do spacewalks. I think most astronauts do. However, I thought everyone would put down spacewalks for their first choice, so I listed spacewalks, or extra-vehicular activity (EVA), second and asked to work in the Shuttle Avionics Integration Laboratory, which we called SAIL.

SAIL was the lab where we tested the software that operated the orbiter and made sure everything worked correctly. It was important work, and I would learn a lot about flying the Shuttle and about operating its systems. I had a solid background with computers going back to Wright-Patterson Air Force Base. I figured that's where I'd be assigned.

But I inadvertently broke the Astronaut Office code. For years astronauts have tried to figure out how to get the assignment they really want. The answer is simple: don't ask for it. I asked to be put in SAIL and instead they put me in EVA where I really wanted to be. I was going to be a spacewalker!

Not only was I going to launch into space and fulfill that dream, but I was also going to leave the orbiter in a suit that would be my own little spaceship, my own cocoon. I would be out in space, free to see the universe all around me. I really felt fortunate because EVA was right in line not only with what I wanted to do but also with my physical attributes and mechanical aptitude.

I was really excited.

Then the swimming thing popped back up to the surface.

Most of the training to become a spacewalker is done in a pool that simulates the weightlessness of space. NASA didn't call their pool a "pool." They called it the Weightless Environment Training Facility, or the tank. It was twenty-five feet deep, seventy-five feet long, and thirty-three feet wide.

To prepare for EVA, we suited up in a training space suit that is pressurized like a balloon. Even though the space suit and crewmember inside it together weigh over three hundred pounds, the suit will float on the surface of the water because of the breathing air that's pumped inside. Lead weights are strategically placed in pockets on belts strapped around the legs and arms of the suit and into other pockets built into the front and back of the suit. Just the right combination of weights makes the suit neutrally buoyant so the suited crewmember doesn't float on the surface or sink to the bottom. Additionally, an astronaut can rotate his or her body with minimal force and will stay where the crewmember wants to be positioned. The conditions are similar to true zero gravity.

Now that I was an astronaut assigned to EVA, the issue was inescapable. To be an EVA crewmember, one has to be certified to wear a space suit in the water for training. To obtain that certification, NASA scuba certification is required. And in order to receive scuba certification, the candidate must be able to swim.

For me it was a catch-22.

I'm sure Mr. Abbey didn't know I couldn't swim when he called to tell me I had been selected to be an astronaut. I couldn't swim at the end of ASCAN training any more than I could at Purdue or in the marsh on Jim Gentleman's farm in northern Indiana.

I talked to the instructors doing the scuba certification, and they told me they would work with me after the rest of the class had been certified. I thought, "Great—after all this work, after all this time, I may not be eligible to do spacewalks because I can't swim."

After the others were certified, the scuba trainers took me to a swimming pool in Clear Lake City near the Johnson Space Center. They thought they'd have me swimming in no time. They didn't. When they tried to teach me to float, I rocked. Literally.

They tried to help me do the basic dead man's float, but even that didn't work. I was beginning to think I was sunk, as far as my dream of spacewalking was concerned.

Several of my 1980 classmates were Red Cross-certified swimming instructors who had taught countless people. Mary Cleave and Bill Fisher were in my class. Mary said she could teach anyone to swim, and Bill had a pool so we went to Bill's house. "This will be a half-hour thing," Mary said. "No problem. We'll have you swimming."

We got in the water. She showed me several strokes, and I followed her example. Each time I eventually sank to the bottom. Sometimes I even went backward before sinking to the bottom, which probably defied the laws of physics.

Mary finally looked me square in the eye and said, "You're right. You can't swim. What you are is a good rock." That was the bad news.

The good news, however, came when the NASA instructors decided that I didn't really need to swim on top of the water for the purposes of NASA scuba certification. I needed to be able to operate under the water. And, hey, that was no problem.

The scuba instructors, Bill Moran and George Price, decided if no one could teach me how to swim, they were just going to make sure I was comfortable in the water and could safely do EVA training.

I wasn't afraid. I just couldn't swim.

Bill and George gave me a snorkel, mask, and fins, and they started me in shallow water. After demonstrating adequate capabilities with the snorkel gear, I graduated to scuba gear. Scuba was fun, and I progressed quickly in the shallow swimming pool. They finally took me to the tank at Johnson Space Center. Bill and George had me go to the bottom of the twenty-five-foot-deep tank, take off all my scuba gear, leave it at the bottom, and surface unaided. Then I had to go back down, which was natural for me, put the gear back on, clear my mask, and return to the top. Once I did that, Bill and George certified me for scuba operations in NASA facilities. I was really relieved!

A major hurdle was overcome, and that was good because performing the "ditch and don" test was all I could muster. In all my time in the Astronaut Office, I never let my scuba certification lapse, because if I had, I would have had to complete that test all over again, which I had absolutely no desire to do!

I've done nine spacewalks, and I held the US records for the number of spacewalks and time on EVA from 1998 until 2007. At this time, nine spacewalks ties me for third most walks performed by any person in the world. Russian Anatoly Solovyev and American Michael Lopez-Alegria have done more. I've spent fifty-eight hours and eighteen minutes "walking" in space and many hundreds of hours more training in tanks.

And I still can't swim.

After I received scuba certification, I still had to be certified to work in the tank in a space suit. I had to demonstrate proficiency in the event of an emergency during training.

Early in the Shuttle program, Shuttle space suits were in very limited supply and were, therefore, reserved for crews assigned to flights. The only suit available for my certification that came anywhere close to fitting me was an A7LB suit from the Skylab era that was made for Pete Conrad. Pete was an astronaut icon. He flew two Gemini missions, on Apollo 12 became the third person to walk on the Moon, and then commanded Skylab 2. Additionally, he was about four inches shorter than me, so we knew Pete's suit would be a tight fit, and it would hurt, maybe a lot. But I had no choice. I needed to get certified so I could do my job.

The A7LB was a one-piece suit with a long zipper that went from the chest to the crotch and up the back of the suit, through which the astronaut entered the suit. Once I got into Pete's suit, it was very snug. When it was inflated, I felt like my shoulders were going to collapse down to my knees. My eyes teared up, but I had to get into the water and get certified.

Underwater, the divers opened the faceplate on my helmet, allowing water to rush into the suit. They placed a scuba mouthpiece into my mouth, and I breathed through the mouthpiece as the divers brought me to the surface. The test only took ten minutes, and I was certified to wear space suits in the water tank. I was happy to get the certification, and I was elated to take off that suit!

Bill Fisher and I were both assigned to work in EVA, and we had a chance to learn from more senior astronauts—Joe Kerwin, Bruce McCandless, Story Musgrave, and Jim Buchli. I was responsible for EVA procedures and tool development, and Bill was responsible for suits. I got to work on the development of many of the tools that spacewalkers would use to repair satellites and to build the International Space Station.

I spent considerable time at the Marshall Space Flight Center in Huntsville, Alabama, working in Marshall's EVA water tank on the early development of on-orbit repair capabilities that were incorporated into the Hubble Space Telescope. That was a real honor because Hubble is without a doubt one of the most well-known successes of the Space Shuttle program.

Launched in 1990, Hubble is allowing human beings to move closer to answering some of the most perplexing questions about the origin of the universe. And at the same time, Hubble is causing astronomers and astrophysicists to rethink other aspects of our understanding of the universe. I worked for a year developing and evaluating spacecraft design features and special tools and techniques to make Hubble an on-orbit repairable satellite. I later used that experience to help support two of the Hubble repair missions in Mission Control when issues developed.

I also did quite a bit of work in the Marshall tank, called the Neutral Buoyancy Simulator, doing other types of EVA development. The Marshall tank was bigger than the tank in Houston. It was a circular tank seventy-five feet in diameter and forty feet deep, and since the Marshall tank was not used to train Shuttle crews, it was easier to schedule for developmental work. I worked on assembly techniques for space structures, developed tools to repair satellites in space, and evaluated potential spacewalking experiments being proposed to NASA by universities and other organizations. It was all fascinating work, and it quickly enhanced my experience and spacewalking skills.

I also got to know many good folks at Marshall who were dedicated to our country's space program. One aerospace engineer who volunteered as a safety diver supporting many of my test runs in the Marshall tank was Homer Hickam. We enjoyed each other's company, but he never told me the stories that he later wrote about in his highly successful book, *Rocket Boys*, which was made into the movie *October Sky*. It is the personal story of his passion for space and launching rockets as a boy. We have a lot in common!

The most senior person in the Astronaut Office when I arrived was John Young. John is one of the most amazing people to ever fly into space. He is also one of my personal heroes. John flew on the first manned Gemini flight with Gus Grissom. Then he flew on Gemini 10 with Michael Collins, who would later serve as Command Module Pilot on Apollo 11—the first mission to land on the Moon. John flew on Apollo 10 with Gene Cernan and Tom Stafford. Apollo 10 did everything Apollo 11 accomplished, except for landing and walking on the lunar surface. Additionally, Captain Young commanded Apollo 16 and walked on the Moon. Later, he commanded the first and ninth Shuttle missions. That totals six missions for John, including two that went to the Moon and one that landed. What a record! For many years, most of us thought NASA would never allow anyone to fly into space more times than John Young.

Captain Young was Chief of the Astronaut Office when I arrived at NASA. John is a real character, with an incredibly good, dry sense of humor. He came to a Test Pilot School graduation while I was at Edwards, gave a talk about the Space Shuttle, and had us all rolling off our chairs laughing. He's one of those guys who says, "Aw shucks," with a deadpan look and makes his words even more hilarious.

Then, on the other hand, John can be hard to get to know, probably because he's so focused. He's a very hard worker. A lot of times when he walked around NASA, he just looked down at his shoes, deep in thought and not paying attention to anyone he passed.

Once when I was a payload officer before I was an astronaut, I was walking up the steps in Building 4. I passed John on the stairs, and he said, "Hi." It made my day. I went home and told Karen, "I met John Young on the stairs today, and he said 'Hi!'"

John commanded the first Shuttle mission STS-1, which launched into space on Sunday, April 12, 1981, at 7:00:03 a.m. Eastern time. STS stands for Space Transportation System. *Columbia* was the first orbiter to go into space, and its flight marked the first time in the history of the US space program that the initial launch of a vehicle had astronauts on board. The most recent American spaceflight before STS-1 was Apollo-Soyuz in July of 1975—nearly six years prior.

The only other astronaut on board *Columbia* was Bob Crippen, the pilot. John and Bob wore pressure suits and sat in ejection seats that could be used to get them out of the vehicle in an emergency; however, the ejection seats were only useful if the emergency happened shortly after liftoff or just before landing. STS-1 would orbit the Earth thirty-seven times, and it was designed to give *Columbia* a good evaluation. The launch had been scheduled for two days earlier, but the first attempt was scrubbed because of a computer problem.

I wasn't at the Kennedy Space Center for the launch. A fellow astronaut, Don Williams, an Indiana native and Purdue graduate, was scheduled to be in Indianapolis that week for a public school presentation, but Don was assigned to do ground support work for STS-1 at the Kennedy Space Center. Launch delays created a schedule conflict, so Don gave me the lengthy speech he had written and asked me to go to Indianapolis in his place. Karen's parents live just north of Indianapolis, so I stayed with them the night before my presentation. I got up early in the morning to watch the scheduled launch and was disappointed when the attempt was scrubbed.

The presentation was going to be taped and then played back to Indianapolis public school students at later times. Students were bused to the recording studio for the videotaping so there would be a live audience. My mother-in-law, Wilma Pearson, went with me.

I hadn't had a moment to read the speech. I was planning to go over it when I arrived at the school system's facility before the taping began, but it turned out I had the only copy, which the studio staff needed so they could retype the speech for the teleprompter. This was the first time I would use a teleprompter. Retyping the speech took so long that there was no time for me to read the speech or practice with the teleprompter. Consequently, the first time I read the talk was from a teleprompter in front of the camera with a live audience and my mother-in-law in the back of the room.

It didn't start out well. I got tongue-tied, and we had to restart the whole thing. There was only time for one more take. The pressure was building.

The second time I was getting along pretty well until I noticed that some letters at the end of each line on the teleprompter were missing. Too many words had been typed per line. Since the missing letters were at the end of the words, I could make out the words easily enough and continue the talk, but then letters disappeared at the beginning of the lines, and that made it a lot harder. I had to figure out what the words at the beginning of each line were and deliver them at the same time.

Somehow I got through that speech. I was wondering what my mother-in-law was thinking. It certainly was not polished, but I think it went pretty well for my first public appearance as an astronaut. I know I'll never forget it.

I was back home from Indianapolis and our family watched TV early on Sunday morning, April 12, as *Columbia* lifted off from the Kennedy Space Center with our hopes, dreams, and prayers. I joined most of the astronauts in our offices at the Johnson Space Center to watch when it landed at Edwards Air Force Base on April 14. Shuttle *Columbia* had only been up two days, six hours, twenty minutes, and fifty-three seconds, but it had been an enormous success. My heart raced as we watched *Columbia* touch down.

When *Columbia*'s wheels stopped after that first flight, NASA personnel rolled a stairway up to the orbiter to retrieve the astronauts. John Young came out first bounding down the steps. His enthusiasm was evident. He walked around the orbiter, front to back, pumping his arms with excitement. He looked like a kid on Christmas morning. Walter Cronkite, broadcasting for CBS News, quipped, "I think the pilot likes his airplane."

He did. The Shuttle era was up and running. Everyone in the Astronaut Office was hooting and hollering. We all liked our airplane!

"This is going to be really good," I thought.

STS-6 was the first time I saw a Shuttle launch in person. It was the first flight of *Challenger*, and I was an escort for the immediate families of the crewmembers. On every flight we had two astronauts who were there to serve as escorts for the families, to assist them in their travel and logistics. But everyone knew the primary reason we were there was to take care of the families in case something went wrong with the launch or landing.

Seeing a Shuttle launch in person was incredible. It wasn't anything like watching a launch on television. It was a physical experience. My whole body vibrated as the sound waves shook us. The speed of the rocket was just awesome, and the sheer energy being released was amazing. The Earth shook, and I could hear the flames crackling. I couldn't wait to be on top of that rocket.

At the end of the mission, I escorted the STS-6 families to Edwards for the landing. During a landing at Edwards or Kennedy, the first thing observers noticed was how fast the Shuttle descended. It dropped to the ground even faster than I drop to the bottom in a swimming pool. The second thing that people noticed was how late the landing gear came down—not until the Shuttle was three hundred feet above the ground. And finally, it was impressive how quiet the landing was. The Shuttle was a 200,000-pound glider when landing. With no engines running, there was almost no sound. The Shuttle went up in a blast that roared in your ears and shook your whole body, but after two sonic booms announced its arrival overhead, it came back to Earth in a whisper.

---•---

KAREN: *Life is really pretty normal for the family of an astronaut. It's like the life of any family that has one parent who travels quite a bit.*

I can remember evenings when Jerry wasn't home. Amy and Scott would get home from school. Scott would ride his bike, Amy would play or read, we would have supper, and then they would do their homework. When I put them to bed, one of them might ask, "Is Daddy coming home tonight?"

Daddy might be working late, doing night flying, or traveling. There were no complaints, but they did want to know when he was coming home. "He won't be home until Thursday," I might answer. But I have to admit that sometimes I didn't remember where he was. One trip blended into another, and sometimes I lost track.

---•---

The STS flight numbering was convoluted. The numbering started out simple enough with STS-1 through STS-9 in sequence. Then the system changed. What would have been STS-10 became STS-41B. I believe that, without saying so, NASA remembered the nearly tragic Apollo 13 mission and decided to avoid having a Shuttle mission with the "unlucky" number. The stated reason for the change was the need to be able to distinguish between flights scheduled to launch from Kennedy Space Center and those that were soon to be launching from the Vandenberg Air Force Base launch site Pad SLC6, which was nearing completion in California.

For my first flight, STS-61B, the 6 indicated the federal government fiscal year in which the flight was scheduled to launch (1986), the

1 indicated it was to launch from Kennedy (a 2 indicated Vandenberg), and the B indicated it was the second flight of the fiscal year scheduled to launch from Kennedy. After the *Challenger* accident and the cancellation of all Shuttle launches planned from Vandenberg, the numbering reverted to the original system with STS-26. Even though the numbering was back to a simpler system, the sequence in which the missions actually flew was far from numerical order. A flight was assigned a number when the mission was planned, and it kept that number even if, due to any number of reasons, the launch order changed.

While waiting for a flight assignment, I had several opportunities to participate in Apollo 13-type ground support efforts to help solve problems that crews encountered in space and to work in Mission Control as the astronaut who communicated by radio with a crew while they were in space. This position in Mission Control is called CAP-COM, for Capsule Communicator, which goes back to the days when astronauts flew in capsules.

In 1984, an electronics box on a satellite named *Solar Max* failed in space. Bruce McCandless and I had recently finished testing a rocket backpack called the Manned Maneuvering Unit (MMU). It enabled astronauts to move around in space untethered. Bruce and I proposed its use for the retrieval and repair of *Solar Max*, and we also started developing the techniques and the tools that would be needed to make the repairs.

NASA approved our proposal and manifested the activities on two flights. I served as the support crewmember for both missions. The first was STS-41B. On that mission, Bruce and Bob Stewart performed EVAs to demonstrate the capabilities of the MMU and test some of the equipment that would be used to capture and repair *Solar Max*. The second mission, STS-41C, was assigned to actually capture, repair, and redeploy the satellite. George "Pinky" Nelson and Jim "Ox" van Hoften were the spacewalkers on that mission.

The STS-41C EVA crewmembers Pinky and Ox completed their second spacewalk, and everything went very well. They had fixed *Solar Max*. The mission was documented in the 1985 IMAX movie *The Dream Is Alive*.

That evening after I finished my shift as CAPCOM, I went back to the Astronaut Office to catch up on paperwork. John Young's office was around the corner. He saw that my light was on and I was working, so he came into my office. John didn't say "Hi." He just marched around my office pumping his arms, just like he did after STS-1, saying, "Boy, you guys did great! That was great stuff you did today! Just great!" And he walked out.

By the time I said, "Thank you, John," he was out the door. It meant the world to me. He wasn't an effusive type of manager, but we admired his unrelenting dedication to our country's space programs.

In November 1984 on STS-51A, spacewalkers Joe Allen and Dale Gardner were assigned to retrieve two satellites that had failed to reach their proper orbit. During flight preparations, I worked with them to develop a plan for the capture. For each satellite Joe or Dale would fly the MMU to the satellite, insert a "stinger" into the satellite's expended rocket nozzle to capture it, and fire the MMU's jets to slow the satellite's rotation. The robotic arm would then grab the "stinger" and place the satellite into the payload bay for the trip home.

On the mission, Joe and Dale were able to insert the "stinger" into the nozzles, but the rest of recovery didn't work the way it was planned. Just days before the mission launched, I had developed a backup procedure in the water tank in case the crew encountered such problems. As CAPCOM, I worked with Joe and Dale to employ the backup procedures to manhandle the first satellite into the payload bay. I did another tank run to tweak the procedures before the second satellite rescue. Both satellites were successfully retrieved and brought back to Earth for repair and relaunch.

In April of 1985, the Shuttle *Discovery* went up with Don Williams as pilot and another Purdue graduate, Charlie Walker, as a payload specialist on STS-51D. They had two satellites to deploy, and there was a problem with one of them. A spring-loaded switch on the side of the communications satellite was supposed to be thrown automatically as the satellite left the payload bay. That switch would have activated the satellite, deployed an antenna, and eventually ignited a booster rocket to push it to its final orbit.

None of those things happened, and it was suspected that the switch might not have been completely thrown. I participated with multiple teams, hectically working around the clock for several days, trying to determine what had happened and how to fix the problem. Finally, we concluded that the only thing we could attempt to do was to try to completely flip the switch by hitting it with "something."

We considered what the crew had available on the orbiter and used those materials on the ground to build two versions of something that looked like a flyswatter. The crew would build the "flyswatters" and do a spacewalk to attach them to the end of the robotic arm. The next day the crew would fly the Shuttle alongside the spinning satellite, move the robotic arm to hold the "flyswatters" against the spinning satellite, and hopefully throw the switch and activate the satellite. It was pretty tricky stuff.

I put on a space suit and went into the water tank to figure out how to attach the "flyswatters" to the robotic arm. I used some of the EVA tools I had helped to develop and found a way to affix the devices. My suggestions had to be acceptable to management, safety, and the EVA and robotic teams.

I was asked to be the CAPCOM during the spacewalk to talk the crew through the EVA procedures, step by step. The crew built the "flyswatters," did the EVA, and attached the devices to the end of the arm. They rendezvoused with the satellite, held the "flyswatters" against the side of the spinning satellite, and caught the switch lever. All of that worked.

We were so happy and congratulated ourselves on a good piece of duct tape engineering, but the celebration was short-lived. The satellite still did not activate. The problem turned out to be in the satellite's wiring. A later Shuttle crew "hot-wired" the satellite, and then it worked as designed. Sometimes Mission Control can't be the hero and save the day, but the whole "flyswatter" episode was still challenging and fun.

By this time I had supported four crews who had performed seven EVAs. I found that each time I helped conduct an EVA from the ground, I became a little more envious of the crewmembers in space. I wanted even more to get up there and outside the Shuttle on a spacewalk myself. Even though I was an astronaut, I hadn't flown in space. I wondered when my time would come.

One day I received a call from Mr. Abbey's secretary asking me to come to his office. As I walked from the Astronaut Office in Building 4 over to Building 1, I hoped this summons might mean a mission assignment.

George Abbey was a hard man to get to know, is impossible to figure out, and he likes it that way. Many times in meetings his eyelids would close, and you would think that he had fallen asleep, but he hadn't. He was listening and thinking. He is a deep thinker, and he is more like a chess player than a checkers player. He thinks many moves, and years, ahead.

Mr. Abbey graduated from the US Naval Academy, but he served in the Air Force as a pilot. It was obvious from the flight crew assignments, especially those early in the Shuttle era, that he favored the Navy. Navy astronauts were assigned to earlier and better flights. The early assignments then rippled into other flight assignments and management positions for those individuals (the chess player at work). In later years, the bias toward Navy morphed to one favoring Marines. But for whatever reason, as an Air Force guy I did okay!

In his typical way, Mr. Abbey invited me into his office and asked about the weather. "Are you keeping busy?" he queried. It was small talk. Then he said, "Well, how would you like to have another job? I know you're busy, but how would you like to do something else?"

"Sure," I said. What else do you say when your boss asks you to do something new?

"Well, you know, we're starting to put this new crew together," he continued.

I practically launched out of my chair.

FIVE

"Liftoff! We have liftoff!"

In 1941, John Gillespie Magee, Jr., an American pilot who flew with the Royal Canadian Air Force during World War II, wrote the poem "High Flight." Most aviators know it and also know that Magee died in a crash at the age of nineteen.

The poem includes these lines: "Oh! I have slipped the surly bonds of earth . . . / And, while with silent lifting mind I've trod / The high untrespassed sanctity of space, / Put out my hand and touched the face of God."

In 1985, after a lifetime of working and waiting, my turn to "slip the surly bonds of earth" finally arrived, and I had great people going with me "to touch the face of God."

My first Shuttle crew was commanded by Test Pilot School classmate and friend Brewster Shaw. Three of my astronaut classmates—Pilot Bryan O'Connor and fellow Mission Specialists Sherwood "Woody"

Spring and Mary Cleave—completed the NASA crew. Two Payload Specialists, fellow Purdue graduate Charlie Walker and Rodolfo Neri-Vela, who was a Mexican citizen, were added later.

Brewster is an exceptional pilot, a very smart engineer, and has proven to be an excellent manager of big programs. Brewster had surprised me very early in our Test Pilot School classes when he leaned over and asked me a very insightful and difficult technical question about our class materials. Until that moment I thought he was just a fighter jock! After Test Pilot School, Brewster was my flight instructor, helping me get my private pilot license at Edwards until he was selected for the astronaut program and departed for Houston. Brewster had flown on the Shuttle once before as a pilot on STS-9.

Bryan was a Marine pilot and test pilot, and he had flown the Harrier vertical takeoff and landing jet. Woody had flown helicopters in Vietnam and was an Army test pilot. One of Mary's claims to fame was her PhD degree in civil engineering—sanitation engineering to be exact. We called her "Sanitary Mary."

Bryan and Woody were the clowns of the crew. Woody has a silly laugh that sounds like Woody Woodpecker would sound if he were a horse. Woody always had a stupid joke to tell, and he is a lot of fun to be around. Bryan has a wonderful dry wit and did a great imitation of comedian Don Novello's character Father Guido Sarducci, who frequently appeared on the television show *Saturday Night Live*.

Charlie Walker was an employee of McDonnell Douglas and was on his third flight, more than the rest of the crew combined. He operated the Continuous Flow Electrophoresis experiment that McDonnell Douglas believed held great promise to manufacture new "wonder drug" pharmaceutical products utilizing the zero-g of space; however, process contamination problems and rapid advances in gene splicing technologies doomed their efforts.

Rodolfo was an electrical engineering professor in Mexico City. He conducted experiments in the mid-deck of the orbiter that were designed and prepared by students in Mexico. He is the only person from his nation to ever fly in space.

We were originally scheduled to launch in January 1985, but as time progressed, our crew was bounced around to several different missions. At one point we were assigned to STS-51L, the disastrous *Challenger* mission. My first flight ultimately would be STS-61B, and we flew in November 1985.

In a fairly unusual but fortunate circumstance, before we launched on STS-61B, I was already assigned to a second mission—STS-62A. Bob "Crip" Crippen, who flew with John Young on STS-1, was going to be the commander on that flight. It was to be the first flight to launch out of Vandenberg Air Force Base. In 1984, I had supported Crip's STS-41C mission. He must have been satisfied with the work I did, because I was assigned to his new crew. My flight with Crip was originally set for January 1986, but we all knew that wasn't realistic, and while I was in orbit in November of 1985, the date of my second mission slipped to July 1986. After waiting five years for my first flight, I was scheduled to have two flights in eight months. The Shuttle program was picking up speed!

In fact, our STS-61B mission was the ninth flight of 1985, and that was the most that NASA would ever launch in a single calendar year. The large number of missions meant all our joint training in the simulator with Mission Control was packed into the last month before the flight. Adding to the pressure, Crip's crew was beginning to train, and I needed to squeeze in some time to work with them. I was worried that because the two flights had moved so close together I might be replaced. I talked to Crip about my concerns. He said, "Don't worry. We'll take care of you." That was great to hear.

STS-61B was the twenty-third flight in the program. We were on *Atlantis*, and the Shuttle preparations for our flight were the quickest in the history of the program—just fifty-four days from the previous launch of *Atlantis* to our launch. That was a fast turnaround. We didn't know it then, but all that rushing was about to come to a sudden, and necessary, halt.

The closer we drew to my first launch date, the more excited I became.

KAREN ROSS: *Preparing for a launch is a very busy time for the whole family. In addition to Jerry training for the mission, there was the opportunity to share the experience with family and friends.*

Planning for a Shuttle launch is like planning a wedding, only in another state on an uncertain date. As in wedding planning, one of the first things to be done was the invitation list. Jerry and I and Amy and Scott compiled a list of names and addresses of people who might want to come to Florida for the launch. The list came from what seemed like our different lives—from Crown Point and Sheridan, from Purdue, from Wright-Patterson, from Edwards, and from Texas. So many people had been important in our lives.

At the Johnson Space Center, NASA arranged a meeting for the crew spouses with the official family escort and the crew secretary. At that first meeting, the family escort gave us a brief orientation and told us what NASA would expect from us. NASA helped the crewmembers with the administration of launch invitations. Our crew secretaries were great, and the crew secretary for each flight and I always became great friends.

While the crew was in training, the spouses selected venues for prelaunch receptions that were held in Florida, usually the day before launch. Some crewmembers had brown bag lunches in parks; others had casual to formal receptions at Brevard Community College, at the Astronaut Hall of Fame, at the NCO or Officers' Club at Patrick Air Force Base, at hotels, and at restaurants, and a few had no reception at all. The spouses reserved the location for a specific date with their fingers crossed that launch would go as scheduled. They selected the food and maybe flowers.

The secretary sent the latest RSVP lists home with the crewmembers, a little like sending a note home with a child to a parent, I always thought to myself. Then the more difficult work began. Each crewmember could have a certain number of car passes for a viewing site on the causeway at Kennedy Space Center, a certain number of seats in the VIP viewing site, and a certain number of seats in the extended family viewing area. The numbers could be different for every flight.

From those who responded saying they could attend, each crew member and spouse identified who would receive the different types of passes. This wasn't as easy as it sounds. For example, Jerry's best friend, Jim Gentleman, his wife, Irene, and their daughter, Christy, would most naturally be on the family bus. But because Jim is a physician, he could be invited as a VIP, giving Jim's family a good viewing opportunity and at the same time making room for three more family members or close friends on the family bus.

When launch dates changed or delays occurred, RSVPs changed. Some people who were planning to come could no longer come, and some people who had not been able to come now were. As launch grew closer, the car pass list, VIP list, and family list sometimes changed on a daily basis.

But, per NASA's schedule, the day finally came for the official invitations to be mailed. Jerry and I wrote three letters: one for guests receiving a car pass, one for guests receiving VIP credentials, and one for guests receiving family bus credentials. The official NASA invitations and launch day instructions for each group were different. The enclosures and address label for each envelope were carefully verified with the latest guest list.

Finally, people at the venue for the reception were given the number of guests we expected, and the food order was finalized.

*And all of this went on while Amy and Scott had school,
homework, and their activities. I was working and keeping
up with things at home while Jerry was in the final stages
of training for a flight. Oh, how hectic! And how exciting!
I would do it again tomorrow!*

———————————●———————————

Normally, about nine months ahead of a flight, the crew dropped everything else they were doing and went into full-time training.

I worked hard and studied hard in preparation for my missions. I wanted to know as much about the orbiter and about our payloads and experiments as I could possibly know. I wanted to thoroughly understand all of the procedures, so if something went wrong, I could help correct the problem. Knowledge was safety, and I wanted to be as safe as possible for the rest of the crew, for myself, and for my family.

———————————●———————————

KAREN: *When November 1 came, I thought, "This is it.
Before this month ends, Jerry will launch." There was no
escaping it. Launch was no longer an abstraction, safely in
the future. It was real, and we were getting closer every
day. With the passing days I learned to think, "Jerry is not
launching today. I won't be nervous today."*

*A week before launch, Jerry packed his bags and the
crew moved into house trailers in a quarantined area inside Building 228 at the Johnson Space Center. A family's
access to the crewmember during the quarantine period for
this flight, like all before the* Challenger *accident, was very
limited. Amy and Scott kissed Jerry good-bye and wished
him a good launch before he left home. They did not see him
again until after the mission. After being checked by the
flight surgeons and receiving "Primary Contact" badges, the
wives of the crewmembers had dinner with them in one of*

the trailers at the Johnson Space Center late the night before we all went to the Kennedy Space Center for launch.

———————————————————●———————————————————

Going into quarantine a week before the flight helped to keep us from contracting any kind of cold or illness that might compromise our capability to launch on time. A very useful by-product of being in quarantine was that it kept what I called the "shirt-tuggers" away from us. There were always people asking us to fly things into space for them or reminding us to not forget this or that. It was distracting. We just needed to get away and focus.

When we arrived at Kennedy Space Center three days before launch, there was a beehive of activity at the launch pad. Technicians were like drone bees, working all over the area, preparing the launch pad and the Space Shuttle. The three communications satellites and the EVA experiments were in the Shuttle's payload bay, and the payload bay doors had already been closed.

———————————————————●———————————————————

KAREN: *The crew flew to Kennedy in NASA T-38s. NASA took the spouses on one of its Gulfstream airplanes. I arranged for a neighbor to take Amy and Scott to the airport in Houston, and Jerry's parents drove from Indiana to Florida in time to meet them when they got off the plane in Orlando.*

That first night the spouses were picked up from our lodgings in Cocoa Beach and Cape Canaveral and shuttled to the Astronaut Crew Quarters quarantine facility for the traditional steak dinner and a brief visit with the crew. At an appointed time, the astronauts escorting us told the spouses it was time to leave, and we were taken back to our hotels. The dinner felt like going to see someone during visiting hours at the hospital—or in jail.

Two days before launch NASA hosted a barbecue for the crew at the Beach House. The Beach House is a small, intimate, formerly privately owned home at Kennedy Space Center that NASA uses for conferences and events. The Beach House has seen a lot of history within its walls going back to the early days of the space program. Each of the crewmembers was allowed four or five adult guests for the barbeque, who were checked by flight surgeons before they arrived.

The Beach House barbecue was always a fun event. The crew was upbeat and ready to go. The guests were proud and so happy to have this time with their husband, wife, son, daughter, brother, sister, or friend. The weather at the Cape was beautiful that day in November. After the guests left, Jerry and I went for a walk on the beach. It was a time of soul-searching and realizing how much we really meant to one another.

The day before the launch, other astronauts called Astronaut Support Personnel, or more familiarly known as "Cape Crusaders," set up the cockpit. They placed all of our cue cards and checklists and the equipment we would use for launch, such as our seats, our communications lines, and our emergency air supplies, in the crew compartment. They set every one of more than one thousand switches and circuit breakers in the correct position for launch and double- and triple-checked them.

KAREN: *The second I woke up on launch day, I thought, "Well, this is it. Before I sleep again we will know if Jerry is going to be safe."*

That morning we had our prelaunch reception at Diamond Ginny's, a Western-themed restaurant in Cape Ca-

naveral. Those were the "Urban Cowboy" days. It seemed like all of our family members and friends were there in Florida. It was a wonderful time to visit with friends we hadn't seen in a long time. And Jerry couldn't be there! Ironically, Jerry had to miss our prelaunch receptions, and they were seven of the nicest times I have had in my life.

———————————————————●———————————————————

On launch day, everyone was ordered from the pad while half a million gallons of liquid oxygen and liquid hydrogen were pumped into the large, orange external tank. When we were awakened four hours and twenty minutes before liftoff, the Space Shuttle was fully fueled and ready for us to climb on board.

I had listened to the detailed debriefings of each of the previous twenty-two crews. When I was at the gym working out or running on the track, I would use that information to vividly daydream about what my flight would be like and what it would feel like riding that huge rocket into space.

But no person and no words can prepare you for the experience of launch. I found out what it's really like on November 26, 1985. We had breakfast in front of the cameras. Then we got into our flight clothes, attended a quick vehicle status and weather briefing, and headed to the pad just two hours and fifty minutes before liftoff.

As we left the crew quarters that afternoon, we waved to the cameras and crowds of well-wishers and climbed into the Astrovan for the twenty-minute ride to the pad. The driver was already on board along with security. The Chief of the Vehicle Integration Test Team, an insertion tech who would help strap us into *Atlantis*, Mr. Abbey, and John Young also rode with us. We drove toward the pad and stopped to let John out. He was going to fly NASA airplanes to check the acceptability of the weather conditions for launch. We drove a little farther and stopped at the Launch Control Center to let Mr. Abbey out.

When we arrived at the pad, it was deserted except for a seven-person closeout crew whose duty it was to put us in the Shuttle, close and seal the hatch, and move to a safe distance at least three and a half miles away.

Walking from the Astrovan to the pad elevator, we sensed that the Space Shuttle was different. Normally, it was just a stack of metal components, but now it was fully powered, the external tank was fueled, and it felt like the Shuttle was alive. It made hissing noises, and it seemed like it was a thoroughbred waiting for the starting gates to open. We rode an elevator to the 195-foot level of the launch pad and waited our turn to get into our seats.

Brewster was in the commander's front left seat on the flight deck. Bryan was in the front right seat. I was immediately behind Bryan, and Mary was in the middle between and behind Brewster and Bryan. Rodolfo, Woody, and Charlie were below in the mid-deck.

KAREN: *That evening was just perfect. The stars sparkled and the air was calm. The wives and children of the crew spent the hours before liftoff in the Launch Director's office inside the Launch Control Center. The secretaries who worked there kept the desk and table covered with sandwiches and snacks and desserts they had made. NASA officials and astronauts flowed in and out wishing us well.*

But these hours in the Launch Director's office were all about marking time. We were waiting. We watched the crew activities on closed-circuit television, and we looked out the slanted windows of the Launch Control Center toward the orbiter sitting on the pad as launch time—T minus 0—grew ever closer.

During the first couple of hours after we were strapped into our seats, things were pretty quiet and slow for the crew. We listened to the launch team chatter on the radio and talked amongst ourselves a bit.

An ergonomist did not create those seats. They weren't designed by anyone who was the least bit concerned about making astronauts waiting for a launch comfortable. They were thin, gray, hard, and designed to secure us as we endured the vibrations and forces of launch. Still, some people, including me, did manage to take a nap while we waited on the launch pad.

Since the time of Alan Shepard, the first American to launch into space, there has been a problem with the call of nature while astronauts wait for launch. When Shepard's flight was delayed on the launch pad, he asked for permission to relieve himself in his suit. That request was initially denied because the engineers were concerned about shorting out electrical leads in the suit. Finally, he was granted permission, and he did what he had to do.

Sitting on the launch pad with my legs positioned above my head for a couple hours, my bladder did get full. But I did have it better than Al. I had a container with the technical name "piddle pack," a plastic bag with a sponge inside, and I did what I had to do. But it wasn't easy with a woman seated beside me.

The planned nine-minute hold, a twenty-minute pause of the countdown clock at T minus 9 minutes, was one last chance for everyone to catch their breath and make sure all was right and that there were no stray airplanes or ships in the restricted area downrange from the launch pad.

It was basically a chance for everyone to do whatever they needed to do to be ready to continue the countdown. When they picked up the count at nine minutes, things moved very quickly.

KAREN: *When we approached the end of the T minus 9 minute hold, on the cue of the escorts everyone got their jackets and cameras. As the countdown resumed, the crew families and our escorts were led down a long hall. We could smell the cornbread and passed the pots of the traditional postlaunch beans simmering on hot plates. We hurried out a door, up a noisy metal stairway, and toward the railing at the edge of the roof.*

At that first launch, Amy (age fourteen), Scott (age thirteen), and I could look down and see our friends and families in the bleachers below. After we got to the roof, the time went so quickly. Our eyes stayed on the launch pad and the Shuttle shining brilliantly white in the special xenon lights that illuminated the pad. We listened to the launch commentary. Being there seemed surreal. Amy and Scott had never been to the Kennedy Space Center, and none of us had personally seen a launch before. "Jerry is out there," I thought. "This is really happening."

———————————————●———————————————

Inside the orbiter as we got closer to launch, Brewster's hands became sweaty. In those days we launched in powder blue cloth flight suits and a helmet that was something like a motorcycle helmet. We were just coming out of the nine-minute hold when Brewster started rubbing his sweaty palms on his pants.

Bryan looked at him and said, "Brewster, I wish you hadn't done that. I was feeling pretty good until you did that!" Then he said, "You know, first John got off the van. Then George got off. Then the guys at the pad strapped us in, closed the door, and they beat it out of here. Do you think they were trying to tell us something?" We all giggled.

Shortly after the countdown resumed, things began to happen. Bryan started the Auxiliary Power Units. When they were activated, we could feel a slight vibration through our seats.

I was very excited. I had a mix of butterflies and anticipation. All those years of hard work were coming to fruition. At the same time, I was focused on my assigned tasks. Just as I did before every flight, whether I was on board or not, I said a silent prayer for the crew's safety and the success of the mission.

As the countdown proceeded, everything happened faster and faster. On later flights, after I had some experience, I told rookies to watch my right leg. "When it starts bouncing—hang on."

During the countdown, "T" stands for time. T minus 31 means thirty-one seconds until liftoff. At T minus 31, control of the countdown was handed off from the Launch Control Center computers to the Shuttle's onboard computers. The onboard computers not only completed the countdown, but they also commanded the entire ascent of the Shuttle into orbit.

T minus 7: Near the pad there was a 290-foot-high water tank. At T minus 7, the valve on the water tank opened, and 300,000 gallons of water cascaded through seven-foot-diameter pipes. I looked back over my shoulder out the overhead windows, and I could see the surge of water begin down at the base of the launch pad. As the water gushed, I decided I'd better turn around and get ready.

By the time the solid rocket boosters ignited, a torrent of water had flooded the mobile launch platform and the flame trench at the base of the pad. The deluge of water protected the pad and the Shuttle from thermal and acoustic damage.

T minus 6.6: The three main engines on the orbiter roared to life, drawing fuel from the 1.6 million pounds of liquid hydrogen and oxygen in the external tank. As the engines roared beneath us, the orbiter "twanged," shifting slightly fore and aft. The engine start got our attention. I could really feel the vibrations, but we were not going anywhere. Not yet. Each of the solid rocket boosters was still attached to the pad by four huge bolts, twenty-eight inches long and three and a half inches in diameter.

On the flight deck where I was sitting, there were six windows in front of the commander and pilot and two windows above Mary and me. We had mirrors on our wrists so we would be able to see out the overhead windows when higher g-forces during ascent wouldn't allow us to turn our bodies and heads.

T minus 0: The solid rocket boosters on the sides of the external tank simultaneously ignited, and the top nuts on the bolts that held the solid rocket boosters to the pad were fractured by explosives at exactly the same instant. When the boosters ignited, it was just like somebody had taken a baseball bat and swung it smartly right into the back of my seat. BAM! Frankly, it felt like the solid rocket boosters had exploded. There was a lot of noise and vibration inside the cockpit.

We were on our way!

The Shuttle leaped off the launch pad and quickly went into the roll program that pointed us in the right direction to get to the desired orbit. But the roll maneuver also caused us to go heads down and uncomfortably slip a couple of inches out of our seats toward the ceiling.

We quickly accelerated away from the pad, and in about forty seconds we were already going faster than the speed of sound. I had thought about what this would be like many times and felt I knew exactly what to expect. But about fifteen seconds after liftoff I caught myself thinking, "Ross! What are you doing here?"

At the same time I was also thinking, "This is really awesome!" The tremendous energy that was being released behind us was putting out such incredible force that I really felt like my body was just being shoved off the surface of the Earth by something really, really powerful.

In fact, that's exactly what was happening! And I sure hoped the Shuttle knew where it was going!

———————————————●———————————————

KAREN: *We heard the words, "10—9—8—7—6." The engines at the bottom of the Shuttle burned bright and the Shuttle swayed. "Oh, God, please keep him safe," I prayed. "5—4—3—2—1—0." The solid rocket boosters at both sides of Atlantis lit like giant Roman candles, and the Shuttle roared and lifted off the pad. It was going. As we watched, it went higher and higher, faster and farther, until the light of the engines arched over the full Moon. There is no other word for it. The launch of STS-61B was beautiful. It was beautiful, and it is still the most perfect launch I have ever seen.*

I turned to Amy, and she was loving it. She was smiling and shouting, "Go! Go! Go!" And I looked at Scott, and he was shaking and crying. I said, "Scott, are you all right?" He replied, "There goes my dad, and he might not come back." Two children, such different reactions! But as we watched, I

*knew Jerry was doing what he had wanted to do his whole
life. His dream had come true. We stood there together and
watched the dot in the sky grow smaller and smaller and
fainter until it disappeared into the black of the night.*

The vibration and shaking were much more intense than I expected,
and the notes that I tried to write on my kneeboard were illegible. The
air noise on the outside of the Shuttle was very loud, much louder than
I had anticipated and much, much more than I had ever experienced
in any commercial or military aircraft. It sounded like it would rip the
paint and tiles right off the orbiter.

At about one minute into the flight, the Shuttle went through maximum dynamic pressure. This was the point during ascent where the
speed of the orbiter and the density of the atmosphere combined to
create the largest aerodynamic forces on the Shuttle stack. The Shuttle's main engines were throttled down some to ensure the loads didn't
get high enough to cause damage to the vehicle. Once past this point,
the Shuttle's main engines increased to full power again. I could really
feel them throttle up.

On my subsequent six launches, I never again sensed the noise
the air made on the outside of the Shuttle was as loud, nor did I ever
again feel that the shaking and vibrations were as dramatic as I did
on this first mission. I don't think the differences were due just to the
fact that I knew what to expect after I'd launched once. I feel that the
differences were attributable to what we wore—the cloth flight suits
and the clamshell, motorcycle-like helmets. After the *Challenger* accident, crewmembers wore, or really it was more like we were encased
in, bright orange launch and entry suits. The launch and entry suit was
a partial pressure suit that reduced the levels of noise and vibrations
that I experienced.

As the Shuttle climbed, the outside atmosphere became thinner,
and the shaking and noise diminished. When the solid rocket motors'
thrust tailed off, it almost felt like we had stopped going up, stopped

accelerating. At two minutes after liftoff, the one million pounds of solid rocket propellant in each of the two solid rocket boosters had been consumed, and the rockets were jettisoned. There were bright orange flashes across the front windows from the smaller rocket firings that pushed the boosters away from the orbiter.

At this point we were already around twenty-five miles high and traveling at four times the speed of sound. The Shuttle's three main engines continued to fire at full thrust. I looked at the displays to make sure everything was working right because it was so smooth and quiet.

As the propellant in the big external tank was consumed, the acceleration increased again. The Shuttle started to bend its trajectory so we were not climbing nearly straight up anymore. We were more horizontal. And then we were really accelerating. At seven and a half minutes after liftoff, the Space Shuttle was accelerating at three g's. This is the equivalent of a car being able to accelerate from zero to almost seventy miles per hour in one second, every second. The acceleration level made it hard to breathe. It felt like a gorilla was sitting on my chest. The Shuttle's main engines were continuously throttled back by the Shuttle's computers so the load level did not exceed three g's. At ten seconds before main engine cutoff, the three main engines were throttled back to their minimum power level.

At approximately eight and a half minutes after liftoff, at precisely the right speed and on the proper flight path, the Shuttle's main engines were shut down. I looked out the windows, but I couldn't see much on the horizon because we had launched at night. We were in space and felt zero-g. At this moment on that first flight, I had the sensation of tumbling head over heels. I felt a little uncomfortable. It was really weird. I looked at the instruments in front of Brewster and Bryan. The orbiter was stable, but I felt like we were tumbling.

Shortly after the main engines stopped, the external tank was released and was later destroyed as it reentered the Earth's atmosphere above the Indian Ocean. At this point we were about one hundred miles high and traveling at almost 17,500 miles per hour, but we were not yet in orbit. Forty minutes later, after one more rocket firing from our Orbital Maneuvering System engines, we were in a circular orbit going around the world every ninety minutes!

The flight plan called for me to get out of my seat quickly and go to work. As soon as I unstrapped and started floating around, I felt fine. No space sickness for me.

As time went on, NASA got better at dealing with space sickness. About 30 to 40 percent of the astronauts have problems with it on orbit. NASA uses promethazine, a drug commonly used in hospitals for motion sickness. If someone is having problems, that person receives a shot in the buttocks at bedtime. It knocks them out, and the next day they normally feel good and are hungry.

When Woody and I got out of our seats after main engine cutoff, we turned on the cameras in the payload bay and opened the orbiter's payload doors.

There was not a lot of time right away to look out the windows. We took off our helmets, removed the emergency air supply from the seats, and stowed them. We detached the mission and payload specialists' seats and stowed them. We activated the toilet and got the kitchen ready.

Living in space on the Shuttle was a lot like going on a camping trip. When we arrived, we had to unpack all our stuff and get everything set up for our stay. Then when it was time to come home, we had to break camp.

The living, working, and floating room inside the Shuttle's crew compartment was small but adequate for the jobs we were assigned to do. Everything on the Shuttle served a practical purpose. It was not luxurious or roomy. The Shuttle flight deck was about six feet tall, twelve feet wide, and eight feet long. The living space on the mid-deck was about seven feet tall, fourteen feet wide, and eight feet long. It was a little crowded, and with seven people on board everyone could be right on top of each other. Literally. You could pass one another left, right, over, or under.

Mary had a problem controlling her body motions in zero-g. By the time we got back home, Sanitary Mary's new nickname was "Oops." She kept bouncing into things and people, and with an impish smile she'd say, "Oops, sorry."

Eating was an important part of the day on the Shuttle, just like at home. We didn't have a galley on my first flight. We had a device that looked like a silver suitcase called the food warmer to heat our food. The galley came along later.

Before a mission the crew went to the Food Lab at Johnson Space Center to taste foods from the NASA menu. Then each of us created menus we wanted in space. A dietician looked at our meals to be sure they met NASA nutrition requirements and suggested changes if they didn't.

Eating on the Shuttle was pretty much the same as it is today on the International Space Station. In the kitchen, crewmembers dispensed water, hot or cold, through a hollow needle into packages of drinks and freeze-dried foods. Other foods were in flexible foil packages, like the Meals, Ready-to-Eat (MREs) used by the military. Some of the foods, including steaks, had been irradiated. We put packages in the food warmer if we wanted the food or drinks to be hot.

At mealtime we could use a meal tray that looked like a customized TV tray. We velcroed our food and beverage packages to the tray and "sat" down for the meal. We squeezed liquid salt and pepper onto our food from small plastic bottles, and our silverware would stay in place on magnets on the meal trays.

We had special straws for the drinks. The straws had clamps on them, and the clamps had to be closed before you took the straw out of your mouth in zero-g. If you didn't, the drink came right up the straw and out the end. Then you had a big ball of liquid sitting on the straw in front of you. If you bumped that ball or moved too fast, shimmering droplets of the drink were sent everywhere. The clamp was one of the things about which we didn't remind rookies. We made them learn the hard way.

Starting with STS-26, which came after my first flight, Karen worked in space food preparation for space program contractors. I loved telling people that since Karen was working full time preparing the space food, the only time I got a "home-cooked" meal was when I flew in space.

KAREN: *I worked with a group of dedicated people who processed flight crew equipment, from towels and batteries to EVA tools and space suits, in a complex of labs and office buildings near the Johnson Space Center. I worked primarily with the Food group. Following NASA specifications, our team purchased, cooked, freeze-dried, tested, and packaged foods to support the space program. We delivered trays of food to simulators when the Shuttle crews were training, stowed trays of food for Shuttle missions, and provided thousands of packages of food for the International Space Station.*

Before each Shuttle flight, the Food group also prepared and served meals for the Shuttle crews while they were in the Astronaut Quarantine Facility at the Johnson Space Center. A few of us took turns traveling to the Kennedy Space Center Astronaut Crew Quarters to lend a hand to the great cooks in the kitchen there and to stow trays of fresh food, which were installed in the Shuttle just hours before launch. When Jerry was Chief of the Vehicle Integration Test Team, he enjoyed being my NASA "boss" at the Astronaut Crew Quarters at Kennedy.

Working in crew quarters was fun and rewarding. By the time the crew was in quarantine, the astronauts were relaxed, or as relaxed as they could be, and anxious to get going. The mood was always light and anticipatory, and we felt privileged to get to support the crewmembers on a more personal basis in those homey settings.

The technicians, engineers, support teams, and managers who processed flight crew equipment for the Shuttle took great pride in their work, and we felt like a family. For me, working in Flight Crew Equipment meant being connected to Jerry's work, knowing more of the people Jerry knew, and having a front-row seat for one of the greatest adventures of our lifetime.

———————————————●———————————————

I was the only astronaut who couldn't complain about the food!

And actually, it was really pretty good. It wasn't Grandma Dillabaugh's apple slice sheet pastry, but I liked it. I liked space potatoes au gratin, hamburger patties, and green beans and mushrooms. The irradiated beefsteaks were good. Later on we had a much wider variety of foods in pouches when they were developed for the International Space Station. I liked those, particularly beef goulash.

I especially liked the lemonade drink. On early flights, the drinking water on the Shuttle had too much iodine in it. It looked like weak ice tea, and it didn't taste good. Lemonade helped hide the taste.

We each selected our own snacks, and there were more in a pantry for all of us to share. All of our menu food was coded with a little colored dot that corresponded to our crew designation. But you had to watch people in space. They would steal your M&M's and cookies.

Never trust an astronaut with the munchies.

We always tried to eat at the same time, so everyone stopped working at mealtime. It was a nice break, and it gave us a chance to catch up on what everyone was doing.

During my first flight, we were in space for Thanksgiving. I missed being home with my family and enjoying all the traditions I love. I also missed the football games. But celebrating Thanksgiving in space was certainly a unique experience, and I had a lot to be thankful for.

We had turkey. We also had a loaf of pumpkin bread. It was a nice meal, and we enjoyed it very much. Some members of the crew took "seats" on the floor of the orbiter as we ate together. Others found a place against the wall.

I ate on the ceiling.

---●---

KAREN: *NASA flew the wives home that night after launch. During the mission, whenever there was something about STS-61B on the news, I ran to the TV and hit the record button on the VCR. The wives were allowed to come to the Mission Control Viewing Room for the spacewalks.*

The Monday after Thanksgiving Amy and Scott went back to school. In Texas, students were only allowed a specific number of absences each semester. In our area, having a parent who was an astronaut was just common enough, and the Shuttle flights had become just routine enough, that our school district did not excuse absences for students whose parent was launching on the Shuttle. Amy and Scott couldn't miss any more days.

---●---

People always ask about the toilet on the Shuttle. It wasn't much. Basically, the toilet was a one-hole outhouse with a vacuum cleaner attached. It used airflow to replace what gravity does on Earth. The urine went through a hose into a wastewater holding tank. When the tank got too full, we dumped the urine overboard and watched the yellow snowstorm outside the orbiter. The fecal matter was aided by airflow down into a bowl where it remained until after the flight when some lucky technician got to remove the toilet and "freshen" it for the next flight.

Astronauts couldn't just sit on the throne; we had to be restrained so we wouldn't float away. We were provided with spring-loaded thigh bars that held us in place. There was a lengthy cue card with step-by-step instructions posted inside the toilet. There were curtains that we velcroed in place when we activated the toilet after we reached orbit. The curtains didn't go all the way to the ceiling or floor, and they didn't provide total privacy. We just gave people their space when they were in there.

It was a suction toilet, but it didn't have good suction. The hole through which the fecal matter passed was not very big. One had to learn to position himself or herself correctly on the seat so there was no mess. At NASA we actually had a training toilet with a camera at the bottom so you could see how to properly line yourself up.

We practiced everything at NASA.

Sleeping on orbit was difficult for me, primarily because I had so much adrenaline flowing. We worked hard, and I didn't get much time to look out the windows during the day, so I stayed up at night. We had sleeping bags that attached to the walls of the orbiter's mid-deck. The bags kept us from floating around at night and hitting people and experiments.

I liked to sleep near a window, so I positioned myself on the flight deck where the view was best. I liked to look at the Earth and take mental snapshots of the world as it passed by. The Earth is so small and precious, and it looks so peaceful and calm from space. I knew that I was very privileged to see the world from this vantage point. I viewed the very thin band that is the atmosphere; it looks so thin and fragile. When you see the atmosphere from space, you more readily understand the need to protect it so that it can continue to protect all life on Earth.

I wish everyone could have the opportunity to see the world from this perspective, especially the world's leaders. Maybe they'd have serious second thoughts about how they treat their people and how they use, or abuse, the world's limited resources. One thing a person can see from space very clearly is that God didn't create any of the borders between nations. They were all established—and fought over—by humans.

In many ways, the Earth looks like it does from a commercial airplane; however, from the Shuttle we could see much farther, and we were traveling much faster. Most Shuttle flights flew at about two hundred nautical miles high. It only took approximately fifteen minutes to go from the West Coast of the United States to the East Coast. If an astronaut didn't have his or her camera in hand, properly set, and ready to take pictures of what the crewmember was looking for, he or she would miss the shot.

On my second flight we passed right over the Midwest. I looked down at St. Louis, Indianapolis, and Lake Michigan. I had the camera ready, and I was able to take pictures of West Lafayette where Purdue is located, northern Indiana where I grew up, and the South Bend area where my sister Judi and her family live.

We did a lot of planned scientific observations from orbit, taking photos of deserts, jungles, lakes, meteor craters, glaciers, and river deltas. The locations were selected by scientists to study the dynamics of changes on the Earth, including those caused by the impact of human activity.

One night we passed over Houston. There are a lot of industrial plants in the area, and there was an incredibly bright flare burning at one of the plants that really caught my attention. As we passed on across the central US, over Detroit, and into Canada, I could still look back at Houston and see that flare.

We did not see any flying saucers, aliens, or UFOs, but I did see several meteorites or pieces of space junk burning up as they entered the Earth's atmosphere underneath us.

On my first mission we had two spacewalks scheduled, but before we got to them we had a lot of other work to complete. Using *Atlantis*'s computers and control panels in the aft flight deck, Woody and I launched three communications satellites from the payload bay.

To launch a satellite, we first remotely opened a sunshade that had thermally protected the satellite until it was ready to launch. We unlocked a turntable that the satellite was mounted on and spun them up to about fifty revolutions per minute (rpm). At exactly the right instant and with the orbiter pointed in precisely the right direction, the computers released and springs pushed the spinning satellite away from *Atlantis*. We fired our Shuttle engines briefly to get a safe distance away, and from there we watched the satellite's own rocket engine ignite and send it to geosynchronous orbit above the Earth's equator. In the geosynchronous orbit, about 22,000 miles above the equator, satellites stay over the same spot on the ground. This is a great advantage for satellites such as communications satellites.

The first day we launched the *Morelos-B* satellite to provide telephone and television service to Mexico. I remembered my grandmother

listening in on that party-line telephone in Crown Point, Indiana, when
I was a kid. Telephone technology and I had both come a long way.

On the second day we launched the *AUSSAT II* satellite for Australia. It was used for maritime and air traffic control, the relay of digital data, telephone communications, and direct satellite-to-home TV.

On the third day we launched the *RCA Satcom K-2* commercial
communications satellite that was used for television broadcasting.

In 1985, communication and data technology was quite different from today. Cellular telephone communication was advancing
but had not yet exploded in the marketplace. In space we used sixteen millimeter film instead of video because film produced better
images than the early video cameras we were provided. There were no
digital cameras. But what we were doing in space helped to launch
the great information technology revolution of the 1990s and early
twenty-first century.

Everything about spaceflight was amazing, and I loved it, but the
thing I really couldn't wait to do was to walk in space. I was like an anxious child waiting for Christmas. Once *Atlantis* was safely in orbit, I
worried that when we checked out our space suits, we would find some
problem. When the suits checked out okay, I started worrying that the
Shuttle might develop a problem that would require us to terminate
the flight before the first EVA. But *Atlantis* performed flawlessly.

One day I'll remember for the rest of my life is the day I performed
my first spacewalk.

The space suit is a sophisticated device made up of numerous pieces.
We call it the Extravehicular Mobility Unit (EMU). Its purpose is to
allow a crewperson to exit the pressurized crew compartment of a spacecraft and enter the vacuum of space, surviving temperature extremes
of plus and minus 250 degrees Fahrenheit. And the suit not only has
to keep the astronaut alive, but it also has to allow him or her to perform the required tasks.

The suit's major components include the lower torso, the hard upper torso, the gloves, the helmet, the primary life support system, and
a secondary oxygen pack. With it on, a crewperson resembles the Mi-
chelin Man.

The lower torso assembly is made of strong cloth and metal parts with a pressure-retaining bladder on the inside. Attached to it are insulated boots and a thermal cover for protection from the extreme temperatures. The lower torso also has bearings that allow legs to bend and hips to swivel. The joints at the knees provide a good range of motion.

The top half of the space suit, the hard upper torso, is made from aluminum and fiberglass, and includes bearings at the shoulders and on the arms at the elbows and wrists to permit easy motion at these joints. These parts also have pressure-retaining bladders inside and thermal-insulating materials on the outside.

The gloves are attached to the ends of the arms and have proven to be the most difficult part of the suit to design and build. The problem is designing them to be comfortable and to minimize fatigue while providing good dexterity and tactility for the spacewalker. The newer versions of the gloves have small heaters in the fingertips. I could have used those for a few snowball fights at Purdue.

Unlike space suits used in previous American space programs, the space suits used on the Shuttle and the International Space Station are not custom-made for each spacewalker. Most of the suit parts have sizing components that can be changed out or adjusted to give the best fit possible. Some of the smallest astronauts, mostly the smaller women, were never able to get a suit that fit properly, so they couldn't do spacewalks.

The helmet attaches to the neck ring of the hard upper torso. It has a sun visor, and lights and TV cameras are mounted on top so the crewmembers can work at night and so people at Mission Control in Houston and inside the orbiter can see what the spacewalker is doing.

The primary life support system looks like a white suitcase attached to the back of the hard upper torso. It contains the oxygen tanks that provide the breathing supply, water tanks that provide water to cool the spacewalker, a battery to power all of the equipment, a radio, a small computer to operate the suit and detect problems, a pump to circulate the cooling water through the suit, a fan to circulate the air, and devices to cool the circulating water, to remove the humidity from the oxygen, and to scrub the exhaled carbon dioxide from the oxygen. It's amazing they can put all of that in one backpack.

The secondary oxygen pack is attached to the bottom of the primary life support system. It is designed to start flowing emergency oxygen to the space suit if a leak should develop. This secondary supply gives the spacewalker thirty minutes to quickly get back inside the Space Shuttle or Space Station. On the front of the hard upper torso is the display and control module. Some of the labels on the module are written backward. The only way a spacewalker can see these labels is with a mirror strapped around the wrist. Using this mirror, the reverse lettering looks correct. These suits are complex and require a lot of training.

Before spacewalkers put the suits on, they pull on diapers and put on Liquid Cooling/Ventilation Garments (LCVGs) that look a lot like long-john underwear with tubes for cooling water sewn into them. Next they put their lower torsos on—and forget what they say about everyone putting their pants on one leg at time. To put these pants on, both legs go in at the same time. Next comes the hard upper torso. It is mounted on a fixture. You crouch underneath it and crawl or squirm your way up and inside while working your arms through the shoulders and down into the arms of the suit. You better not be claustrophobic!

Electric cables and cooling and ventilation lines are connected. The lower torso assembly is brought up, attached to the upper torso assembly, and locked in place. The suit was so snug that I could not inhale a complete chestful of air.

With all this accomplished, next come comfort gloves and wristlets that are used to absorb perspiration and minimize the abrasions that the fingers and wrists will experience during an EVA. Then the gloves are attached and locked in place. Next the oxygen flow to the suit is initiated, the fan and pump are turned on, the helmet is locked in place, and a procedure is started to purge the nitrogen from the suit. This is an important step in the process. Just as a scuba diver has to follow a prescribed protocol to prevent decompression sickness, spacewalkers follow a procedure that removes most of the nitrogen from the bloodstream, the muscles, and other soft tissue in the body before they start their spacewalk. This procedure requires the spacewalkers to breathe pure oxygen for up to four hours before they can depressurize the airlock and exit into the vacuum of space.

During spacewalks the suit is pressurized to 4.3 pounds per square inch (psi) with pure oxygen. This pressure was selected as a compromise between a higher pressure to help lower the probability of decompression sickness and a lower pressure to make the suit less stiff and less fatiguing to work in. Even at the relatively low pressure of 4.3 psi, wearing the rigid, uncomfortable suit is arduous. The gloves are stiff, and every time an astronaut opens and closes his hand it is like squeezing a stiff rubber ball. EVAs normally last six hours or more. Have you ever tried squeezing a rubber ball repetitively for that long?

Once inside the space suit, the wearer mounts the tool carrier, tools, and tethers required for the planned spacewalking task onto the suit. We always have a steel cable safety tether connecting us to the Space Shuttle or Space Station to make sure we don't go floating off into space, lost forever. We also have tethers on all the tools and equipment we carry outside to make sure they don't float away. All of the tethering and untethering adds to the fatigue that the hands experience during a spacewalk.

Everything I had gone through to reach this moment was well worth it. When I finally opened the hatch for the first EVA and stuck my helmet into the payload bay, I wanted to let out a great big shout of elation. The only reason I didn't was my fear that the rest of the crew and the flight controllers on the ground might think that Ross had finally lost it and order me to get back inside at once.

When I floated through the hatch, an amazing view burst before my eyes. I literally moved in an instant from the cramped airlock into infinity. The large visor that covers an astronaut's face gave me an ear-to-ear view that was like standing in the middle of a huge bay window. There were no structures or atmosphere to obscure my view. It was crystal clear. I was face-to-face with the universe.

During our two spacewalks, Woody and I performed tasks to determine the feasibility of manually assembling structures in space. We used what were called the EASE and ACCESS experimental hardware to perform these evaluations.

The EASE experiment was designed by the faculty and students at MIT. When completed, EASE looked like a pyramid standing on its point. It was comprised of six, twelve-foot long, six-inch diameter aluminum beams and four nodes, or couplings. The assembly required one of us to free float at the top of the assembly, receive the beams from the other crewmember, maneuver the beam into position and alignment, and then attach it to the structure, while still maintaining our position. The free-floating task was not the easiest thing to do, and it certainly became fatiguing to our hands and forearms as we repeated the process multiple times.

The second assembly experiment, ACCESS, was a product of the NASA Langley Research Center in Virginia. Doug Heard was the lead engineer. He and his team had experimented for several years in the water tank at the NASA Marshall Space Flight Center in Huntsville, Alabama, to refine their design. I had participated as a test subject in some of those activities and was thrilled to be involved in the actual testing of their hardware in space.

Their design resulted in a triangular cross-section truss that could be built to be as long as desired. We had enough components to complete a forty-five-foot structure. Both Woody and I stood in foot restraints, which held us in place and freed our hands for work. The truss assembly was built on a lazy Susan-like assembly jig that we rotated as we completed each assembly task in sequence.

Sometimes during the EVAs, Woody or I was attached to the end of the robotic arm in a foot restraint and Mary operated the arm to move us to work sites. Mary had spent countless hours on Earth practicing control of the arm. She said she was glad to finally be moving around real people instead of the "cardboard astronauts" used in simulations at home.

Like all patriotic construction crews, we attached a US flag on our completed metal ACCESS tower and saluted it. We also brought out a sign advertising ourselves as the Ace Construction Company. Our motto was, "We will build anything, anytime, anywhere."

We simulated installation of an electric cable on ACCESS, and all of our activities were performed to assess the feasibility of eventually building a permanent structure in space. At the end of the second EVA, Woody and I said, "Let's go build a space station!"

Our own body weight plus the space suit totaled about four hundred to four hundred fifty pounds. The EASE beams we lifted weighed about sixty-four pounds. Although it was easy to lift everything in space—one person could lift the entire forty-five-foot tower—our hands became incredibly fatigued. When we got back inside, Woody said his hands were so tired that it was all he could do to close the hatch.

I was tired, too. My LCVG was soaked with perspiration. But the part of my body that was most tired was my brain. While on a space-walk, my mind was going a million miles an hour. I was constantly thinking about what I was supposed to do next, double-checking what I was doing, checking my safety tethers, communicating with the rest of the crew and the ground, and thinking about what my EVA part-ner was doing. Everything had to be done very carefully; I couldn't let my thoughts wander. So while physically I was tired, mentally I was exhausted.

I might have been tired, but I was in no hurry for our return to Earth. I was having too much fun. However, after six days and twenty-one hours, our journey ended at Edwards Air Force Base.

The preparations for the entry and landing started many hours be-fore the actual landing. The checkout of all of the Shuttle's flight control systems and attitude control rocket engines and much of the reconfigu-ration of the crew compartment were accomplished the day before. The remainder was finished after getting up on landing day. We had to tear down our on-orbit "camp" and turn the crew compartment back into an aircraft cockpit. We stowed exercise equipment, mounted our seats, and set up our checklists and cue cards. During reentry, the tempera-ture in the crew compartment increased due in part to the high tem-peratures on the outside of the orbiter as we slammed back into the Earth's atmosphere at nearly five miles per second. To compensate, as soon as we got up on landing morning, we set the cabin temperature control to a colder temperature.

Brewster and Bryan verified the navigational system's accuracy to ensure a safe arrival at our intended landing site. We closed the giant payload bay doors, turned off the TV cameras, and stowed any remain-ing equipment. We got into our flight suits and strapped ourselves into our seats. We then connected the air hose and communications leads to our helmets and put them on. Brewster and Bryan pointed the tail of the orbiter—not the nose—in the direction we were flying.

At the appropriate time, about one hour before landing and about halfway around the world from where we planned to set down, we fired our rocket engines and started the entry process. We only needed to slow the orbiter by about four hundred to five hundred feet per second of the approximately twenty-six thousand feet per second we were traveling to cause the Shuttle to descend into the Earth's atmosphere. After completing the deorbit rocket firing, the Shuttle was flipped around so that its nose was pointed in the direction of travel. We were then oriented properly so that the thermal tiles on the outside would protect the Shuttle's aluminum structures.

About one-half hour after firing the deorbit rocket engines, we slowly began to feel the early effects of the Earth's atmosphere on the Shuttle. Objects no longer floated. We started to sink into our seats, and we had to hold on to our checklists or they would fall to the floor. By the time we reached 0.5 g, we felt like it was more than 1 g. We felt very heavy, and when we reached higher loads, up to 2 or 2.5 g's, our bodies strained against the forces and we sweated heavily.

As we flew lower into the atmosphere, the friction of the air slowed the Shuttle and created temperatures of more than 2,500 degrees Fahrenheit on the exterior of the vehicle. The protective tiles were earning their keep.

We saw a beautiful display of colors out the windows as ionized plasma gases glowed. First the colors were pastels—oranges, pinks, and greens. They swirled in the corners of the front windows. As we descended and the outside temperatures rose even higher, the colors became brighter and changed to lemon yellow and bright white and filled the view out the front windows. Looking out the overhead windows, I could see what at first looked like a flickering flame. In actuality it was the plasma gas flowing around the exterior of the orbiter and behind us like the wake of a boat in water. This display grew into a continuous tube of green plasma swirling behind the Shuttle, like a neon tube being pulled across the blackness of space.

Me and my grandfather Ellis Dillabaugh outside the workshop of his house moving business. (*Phyllis Ross*)

Grandpa Joe and me with my dog, Prancer. (*Phyllis Ross*)

My best friend, Jim Gentleman (left), and me showing my little sister Janet our new bikes. (*Phyllis Ross*)

Our family in 1958. L-R: Janet, Mom, me, Dad, and Judi. (*Donald Ross*)

Dad and I ready to go to the ball game. Behind us across the field is Grandma and Grandpa Joe's house. (*Phyllis Ross*)

My Honda 90 motorcycle, June 1965. (*Jerry Ross*)

Holding the goalpost taken from Ross-Ade Stadium after Purdue defeated Indiana University in 1966, sending Purdue to the Rose Bowl for the first time L-R: Circle Pines house brothers Max Lortie and Roger Eacret and me. (*Jerry Ross*)

Karen and I were married on January 25, 1970, in Boxley United Methodist Church. (*Morris Pearson*)

A ramjet engine test firing at Purdue University's Jet Propulsion Center. During my master's program, I worked on this project with my major professor, Dr Cecil Warner. (*Jerry Ross*)

A ramjet missile test I conducted on the Holloman Air Force Base, New Mexico, sled track, February 16, 1973. The sled and ramjet reached a top speed

Our family at Edwards Air Force Base in 1976. L-R: Amy, Karen, me, and Scott. (*Karen Ross*)

I received the USAF Test Pilot School Outstanding Flight Test Engineer Award, Class 75B. Former astronaut and the Edwards Air Force Base Flight Test Center Director, Major General Tom Stafford, presented the award. (*Jerry Ross*)

I flew on a total of twenty-three flight test missions on the B-1A bomber at Edwards Air Force Base. (*Jerry Ross*)

Amy and Scott get ready to launch a model rocket at Edwards Air Force Base. (*Jerry Ross*)

I took this picture from a NASA T-38 just as STS-4 *Columbia* touched down at Edwards Air Force Base on July 4, 1982. (*NASA*)

Meeting celebrities is a wonderful perk of being an astronaut. After the STS-4 landing, Guy Gardner and I met Roy Rogers, one of our childhood heroes. (*NASA*)

In 1985, Woody Spring (not seen) and I performed two spacewalks (EVAs) on STS-61B to demonstrate the feasibility of assembling structures in space. (*NASA*)

Mrs. Effie Laney, my fourth grade teacher. After my first Space Shuttle mission, I thanked her for the important influence she had on my becoming an astronaut. (*Phyllis Ross*)

A die-hard Cubs fan, Dad was in heaven the day I returned the Cubs banner I had flown in space. Here we are in the Cubs dugout at Wrigley Field. L-R: Brother-n-law Bill Rattazzi, sister Jan Rattazzi, Karen, me, Mom, Dad. (*Donald Ross*)

At liftoff on STS-27, the Space Shuttle's three main engines and two large, white solid rocket boosters combined to generate over 6.5 million pounds of thrust. It was a great ride! (*NASA*)

After STS-27, Director of the CIA Judge William Webster presented me with the National Intelligence Medal of Achievement. (*CIA*)

Visiting the cottage, at the Baikonur Cosmodrome in Kazakhstan, where Yuri Gagarin slept the night before his historic spaceflight. (*Jerry Ross*)

Me getting into the Russian space suit through its back hatch at the Star City Cosmonaut Training Center northeast of Moscow. (*Jerry Ross*)

A big smile after we saved the $670 million Compton Gamma Ray Observatory by manually freeing a stuck antenna during this spacewalk. This is "The Picture"! (*NASA*)

The STS-37 crew enjoying a lighter moment with President George H. W. Bush during our visit to the Oval Office. L-R: Ken Cameron, me, Steve Nagel, President Bush, Linda Godwin, and Jay Apt. (*Official White House Photograph*)

Our STS-55 crew gag picture, featuring the German crewmembers in Western attire and the American crewmembers in lederhosen. Our three secretaries served as our biergarten barmaids. (*Jerry Ross*)

Conferring with the legendary Gemini and Apollo Flight Director Gene Kranz (left) and former Space Shuttle Flight Director Randy Stone (middle) in Mission Control. (*NASA*)

Spending time with my family before the STS-74 launch at the Kennedy Space Center's Beach House. L-R: Daughter-in-law Faith, Karen, Amy, me, Scott, and Mom. (*Karen Ross*)

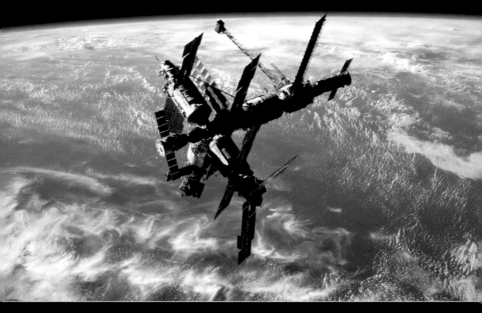

A departing view of the Russian Mir space station after our STS-74 visit. (*NASA*)

Karen and me in the launch pad's White Room just outside the hatch of the Space Shuttle *Endeavour* prior to STS-88. (*NASA*)

The infant International Space Station after the first assembly mission as seen by our STS-88 crew aboard the departing Space Shuttle *Endeavour*. (*NASA*)

The "Spacewalker" in my "office," surrounded by the International Space Station and the Space Shuttle *Atlantis.* Can you find me? (*NASA*)

I was the first person to launch into space seven times. Mike Bloomfield, the STS-110 commander, presented this banner with my seven mission patches to me in the International Space Station's US *Destiny* laboratory. (*NASA*)

Most of the STS-110 crew relaxing on the International Space Station with the two American ISS crewmembers Carl Walz, with the keyboard, and Dan Bursch, far right. (*NASA*)

...had a great view from the end of the Space Station's robotic arm during STS-110 on my last spacewalk. (*NASA*)

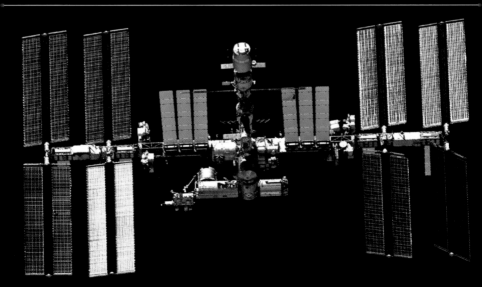

A beautiful view of the nearly complete International Space Station. (*NASA*)

At Purdue University for the National Football Foundation Honors dinner on June 16, 2011. I'm standing in the back row (far right) with legendary Boilermaker football players (L-R) Kyle Adams, Leroy Keyes, and Drew Brees. Seated in front of us are (L-R) Neil Armstrong and Gene Cernan. (*Purdue University*)

The traditional astronaut poker designed to use up the commander's bad luck he must lose before the crew can depart for the launch pad. I am dealing the cards to the crew of STS-135, the final Shuttle mission. (*NASA*)

The students and staff at the Jerry Ross Elementary School, Crown Point, Indiana. I am very proud of all of them! This picture was taken in the fall of 2011 with beautiful Indiana fall colors behind the school.

(Courtesy of Stephen Martin Photography)

A sunset as seen from the International Space Station. (*NASA*)

Our family. Front, L-R: Granddaughter Emily, Karen, me, granddaughter Katie; Back, L-R: Daughter Amy, son Scott, Scott's wife Faith, granddaughter Cassidy. (*Jerry Ross*)

The computers steered the Shuttle to maintain the tiles within their thermal limits, to control structural loads, and to make sure we arrived at the intended place and at the proper altitude and airspeed for the commander and pilot to be able to take manual control and land.

Just after the Shuttle decelerated below the speed of sound, Brewster took control. He followed a computer-prescribed curving path to align the orbiter with the runway. The orbiter descended at a steep angle of about twenty degrees with respect to the ground. This is eight to ten times steeper than a commercial airliner approaches the runway.

We were flying at three hundred miles per hour. As we neared the ground, Brewster decreased the descent rate, and as a result the airspeed was reduced. At three hundred feet above the ground, Bryan deployed the landing gear. Shortly thereafter the Shuttle touched down on the runway at about two hundred miles per hour. The nose wheel hit the runway with a thunk, and Brewster used wheel braking to bring *Atlantis* to a stop about 10,700 feet from where it first touched down.

We turned off systems that would be hazardous to the ground technicians and followed our checklists to configure the Shuttle for our exit. I unstrapped and tried to stand up. I hardly budged. I verified all of my straps were disconnected and tried harder to stand up. It felt like someone had glued my pants to the seat, but I made it.

We all felt heavy, and we were sweating due to the physical exertion it took to operate in Earth's gravity after seven days in zero-g. It was hard to believe that it was so natural to live on the ground only seven days before. Fortunately, this feeling only lasted about thirty minutes.

KAREN: Atlantis *landed on the Tuesday after Thanksgiving. NASA flew the spouses to Edwards in California, and we stood in a structure that looked like a scaffold to watch the landing. We searched the skies for a white dot or for a glint of reflection. Nothing. We were told it should be* *right there. Then we heard two rapid-fire sonic booms and someone spotted the Shuttle. "There it is!" Fingers pointed into the sky to direct us to the right place to look. One by one, each of us picked it up. Then we watched the Shuttle glide in like a very large white bird. There was no sound but a swish as it landed.*

It was a long wait before the crew was cleared to exit the orbiter. But then there Jerry was, safe and sound and more excited about space travel than ever!

———————————————— ● ————————————————

After exiting the Shuttle, we went to a medical facility at Edwards for our postflight exams. After my physical, I was cleared to get a shower. I walked out of the doctor's office, down a hallway, attempted to turn into a bathroom, and walked smack into the doorjamb. I hit it so hard that I bounced off. I quickly looked both ways. Thankfully, no one had seen me.

———————————————— ● ————————————————

KAREN: *We all flew back home to Houston, and the NASA community and friends of the space program were waiting on the tarmac at Ellington Field, as they did for the return of every crew throughout the Shuttle program. The crew-members said a few words to share some of their experiences and to thank those who had helped make the mission possible. Then Jerry and I got into our little brown pickup, and Jerry drove us home.*

———————————————— ● ————————————————

The mission lasted not quite seven days, and I was already back in my home about ten hours after the landing. I was sitting in my rocking chair talking with Mom, Dad, Karen, Amy, and Scott about the incredible things I had just experienced when the doorbell rang. It was the most natural thing in the world for me to want to just push off from my chair and float over to the door to see who was there. I caught myself in the act of positioning myself for the push-off, remembering that floating wouldn't work down here. I smiled at myself, got up, and walked over to the door on my own two feet to greet our visitors. After flights I have watched several of my crewmates release something from their hand and expect it to stay there floating in the air. I have helped clean up more than one mess, and I may have created one or two myself. The human body really likes and readily adapts to zero-g!

KAREN: *Jerry told Phyllis, Barney, Amy, Scott, and me as much as he could about the flight before he just had to go to bed after a day that had started for him nearly twenty-four hours earlier in space.*

The next morning he went back to work. He was excited to see the film from the flight, which was already being developed. That night, one week late, Jerry, his parents, Amy, Scott, and I sat down to a turkey dinner and celebrated Thanksgiving. We had a lot to thank God for.

I experienced an interesting sensation during the first couple of nights after most of my flights. When I was in bed those first few nights, I had the feeling that I was not lying on the bed, but rather floating just above the mattress. I knew I was not floating, and I was not imagining that I was back in zero-g. I believe the sensation was a result of the "touch sensor" nerves in my body becoming desensitized and not sending out as strong a sensation as they had before the flight. I also think that, along with vestibular disturbances and muscular deconditioning, this touch sensor desensitization affects a crewmember's ability to maintain balance and walk for a short period after returning to Earth, even after a short spaceflight.

Since I have flown seven Space Shuttle flights, each with significantly different mission content, I am frequently asked which flight is my favorite. I think that is much like asking a mother which of her seven children is her favorite. However, I guess if I were pressed to pick one, I would pick STS-61B, primarily because it was my first flight and because I got to do my first two spacewalks.

After STS-61B, I had a wonderful sense of fulfillment. I had reached my lifelong goal, and I had the satisfaction of a job well done. I was filled with an almost overwhelming combination of feelings of contentment, satisfaction, accomplishment, and having "reached the top of the mountain." I had worked, studied, with God's help made some good decisions, earned some good breaks, and been just flat lucky. And as a result, I had flown in space and completed two spacewalks.

Now, there was only one more thing I wanted to do.

I wanted to do it all over again!

SIX

"Obviously a major malfunction"

I believe in the power of prayer. I believe we should pray that God's will be done, not ours. We should pray for help in understanding God's plan and what He knows is best for us.

So it's appropriate that my second flight began with a group prayer.

If you have ever wondered what went on in the Astrovan as the astronauts took their final ride on Earth before launching into space, you might be surprised at the stories. We were excited about what we were doing and anxious to get started, but there was also some unspoken tension. We knew what could happen.

On my second flight, STS-27 in 1988, Dan Brandenstein, then Chief of the Astronaut Office, joined us for the ride. The Astrovan stopped at a prearranged point so he could get off to fly NASA airplanes to check the weather. As Dan stood up to leave, he said he'd like to lead us in a prayer. I thought that was really nice, so I bowed my head, closed my eyes, and listened for his comforting words. I can always use a prayer, and I especially welcomed one at that moment.

Dan paused and said, "God help you if you screw up. Amen." And that was it. He left. We all roared with laughter. Of course, we also knew he meant it.

After STS-61B, I wanted to go into space again—and the sooner the better. Fortunately, that was in the works. I had already been assigned in February 1985 to my second Shuttle mission, nine months before I flew my first one.

My second flight, STS-62A, was to be a unique opportunity to fly on the initial Shuttle mission scheduled to launch from Vandenberg Air Force Base on the coast of California. The flight was going to launch into an orbit that was inclined at an angle of seventy-two degrees with respect to the Earth's equator. That was to be by far the highest inclination ever flown, and we would pass over parts of the Earth closer to the poles than any previous human spaceflight. We were going to be able to see things that no one had ever seen before from space.

The mission was a military flight, but most aspects of what we were going to do would not be classified.

There were two primary objectives. One was to deploy an experimental military satellite. The second was to conduct a series of observations using special telescopes and instruments mounted in the payload bay of the orbiter. The observations would gather infrared, ultraviolet, and visible light spectrum scientific data on natural and induced phenomena in the very upper reaches of the Earth's atmosphere. The information would be used in the development of President Ronald Reagan's "Star Wars" ballistic missile defense systems.

I started training in earnest for this flight in January 1986. On January 28, our crew was at Los Alamos National Laboratory in New Mexico. We were being trained on the equipment the Laboratory's scientists had built for our mission.

We had a television on in the corner of the room with the sound turned down, and we periodically checked the progress of the launch countdown for STS-51L—*Challenger*—at the Kennedy Space Center. That flight was scheduled to deploy a NASA communications satellite, and it had a small payload on board to observe Halley's Comet. The public was largely following this launch because the crew included teacher Christa McAuliffe, who was planning to teach classes for kids from space.

The Shuttles had been doing a lot of flying. My first mission, STS-61B, landed in early December of 1985. STS-61C launched on January 12, 1986, but only after seven delays. On board that flight was US Congressman Bill Nelson of Florida, the second sitting politician to fly in space. The first had been Senator Jake Garn of Utah, in April of 1985.

STS-61C was scheduled to land on January 17, but that day interfered with the preparations at Kennedy Space Center for the next scheduled launch—*Challenger*. So NASA tried to bring 61C home on the January 16; but, due to bad weather, it ended up landing at Edwards Air Force base on January 18.

This impacted the schedule. *Challenger* had originally been scheduled to launch on January 22, but it was pushed back to January 23 and then January 24 with the landing delays. Weather problems at Kennedy Space Center delayed the launch once again to January 25. More bad weather pushed the launch back to January 27, and then problems with an exterior access door panel moved the launch to 11:38 a.m. EST on January 28.

The temperature for January 28 was expected to be below freezing, the coldest ever for a Shuttle launch. Reporters waiting around at the Kennedy Space Center for *Challenger* to launch were giving NASA officials heat about the many delays. There was a lot of internal pressure on everyone to launch and meet our aggressive flight schedule.

Challenger finally launched on January 28. Sixty-eight seconds later, CAPCOM Dick Covey, another Purdue graduate, told the crew, "*Challenger*, go at throttle up."

"Roger, go at throttle up," *Challenger* Commander Dick Scobee responded.

That was the last communication from *Challenger*, although there were reports that Mike Smith, the pilot, was heard to say, "Uh oh!" *Challenger* came apart at seventy-three seconds. There was a long silence, and then Public Affairs Officer Steve Nesbitt said what everyone knew, "Obviously a major malfunction."

It was just after 9:30 a.m. in New Mexico. As we worked, Dale Gardner, a member of our crew, was facing the TV, and he said, "What's that?" We all turned and looked in stunned silence; everyone had a different idea about what they were seeing. I saw debris falling out of the sky and the solid rocket boosters corkscrewing. My first reaction was that they were doing a return to launch site abort. But we were all very knowledgeable about Shuttle launches. We knew what we were seeing was very wrong. We just didn't want to believe it.

We stopped our work and made arrangements to immediately fly back to Houston. Pete Aldridge was a payload specialist on our crew. He was Under Secretary of the US Air Force, Director of the National Reconnaissance Office, and he would soon be promoted to Secretary of the US Air Force. He had an Air Force business jet waiting for him in Albuquerque. Most of the crew flew back in T-38s, but I had a bad head cold, so I opted to fly to Houston with Pete in his pressurized plane. I remember the flight vividly. It was very quiet, and I stared blankly out the window. I prayed for the crew and their families.

We were all depressed and in a state of shock, trying to learn everything we could and still holding out a faint hope for the crew's survival. Once we got back to Houston, our worst fears were confirmed, and I headed home to be with my family.

Karen and I went to Ellington Field late that night to show our support for the families when they arrived back in Houston. We visited most of their homes over the next week and offered to help in any way we could.

I wasn't assigned to any jobs supporting the accident investigation, so I decided I would help support the families by attending as many of the services and memorials as I could. It was painful, but I hoped it helped.

The families received cards and letters from thousands of people all around the country, and they decided they would answer every one of them. Karen helped with that, and her mother, Wilma, came for a visit and helped, too.

KAREN ROSS: *I was nervous the whole month before Jerry's first flight. When the mission was over and I knew everything was all right, I thought, "What a shame I can't go back and enjoy it."*

Then I watched my friends, the spouses of the Challenger *crew, deal with their loss. The NASA community felt broken and rallied around them. There were tears, shared memories, and even some laughter. The families showed so much grace and strength.*

I learned from my first flight experience and from Challenger. *For the rest of Jerry's career, I tried not to be anxious and to just enjoy everything leading up to the flight. That period of time is so special, and it goes so fast.*

If something did go wrong, there would be a lifetime to deal with the sorrow.

I knew all seven astronauts we lost that day, and I knew most of them very well.

Dick Scobee: Dick was commander of the *Challenger* mission. He was a little older than most of his classmates when he was selected by NASA in 1978. I always viewed him as "the Wise One." He had a calming influence in any discussion and carefully considered any words before he spoke them, but he also liked to laugh and have fun. Dick and I had been stationed together at Edwards Air Force Base and casually knew each other there. At Edwards, Dick had a high-visibility assignment testing the experimental X-24B lifting body vehicle. I supported Dick's first Space Shuttle flight, STS-41C, and thoroughly enjoyed working with him during that mission. Dick was a perfect choice for the commander of the STS-51L flight, and he treated his crew as an extension of his family. His wife, June, is a dedicated educator, and Dick was committed to making sure that Christa McAuliffe's Teacher in Space Project would be a big success for the students all across America.

Mike Smith: Mike was the only *Challenger* crewmember from my astronaut class. He was the pilot on *Challenger*, and he was a class act. He also liked to laugh and have fun. He was soft-spoken, a little on the quiet side, and a great family man. Mike was a US Naval Academy graduate and had flown in Vietnam. Although still a fairly junior member of the NASA team, he had already been given a couple of leadership roles in the Flight Crew Operations Directorate. He was well respected by his managers and his contemporaries and had a very bright future with NASA.

Judy Resnik: Judy was smart and attractive. She achieved a perfect SAT score at the end of high school and was an accomplished pianist. She earned her PhD in electrical engineering from the University of Maryland. She was fun-loving and a jokester. I worked with Judy on the development of procedures and checklists for the operation of the Canadian-built robotic arm that was carried in the payload bay on many of the Space Shuttle missions. I was a CAPCOM on her first Space Shuttle mission, STS-41D. During that flight, I was the planning shift CAPCOM and served mostly during the crew's sleep period. On two consecutive nights during the mission I had to awaken the crew, a highly unusual thing to do. The first night, there was a measurement that was only one increment away from causing an alarm to sound in the orbiter, which would have rudely awakened the crew. Mission Control decided to have me call them and ask them to adjust the alarm limit before it was tripped. The next night, I had to awaken them again and ask them to reconfigure some switches when we determined that there was an oxygen leak in the Space Shuttle plumbing. After the crew was back on the ground, I met Judy one day at the door of the Astronaut Office mailroom. She surprised me when she jokingly grabbed my shirt with both fists and said, "Ross, the next time you wake me up in the middle of the night, it better be with your elbow!" That was Judy!

Ellison Onizuka: El and I knew each other very well from Edwards, where he was on the Test Pilot School staff when I was a student. We played softball together at Edwards, and we traveled together to Houston in the fall of 1977 when we were both interviewed for the 1978 class of astronauts. El was selected, and I wasn't. We played on the Astronaut Office softball teams for several years until his death. Our children would play together at the park while we were playing softball. El was always ready to have a good time and play a joke or two. He was a little shorter than me and very strong. He told me he had developed much of his strength from carrying sacks of coffee beans down the slopes of the hills of Kona, his home island in Hawaii.

Ron McNair: I first got to know Ron when we were flying in T-38s, training to chase the Space Shuttle during the landings of STS-3 and STS-4. Ron and I flew in the backseats of the T-38s and assisted the pilots in locating the descending Shuttle and monitored the T-38 systems while the pilots concentrated on the rendezvous with the Shuttle. Ron and I used our Hasselblad cameras to completely document the condition of the thermal protection tiles on the undersurface of the orbiter prior to the landing. Ron was on the more serious side but still liked to have fun. While we were spending some time in El Paso, Texas, training to support the STS-3 landing, he joined a band impromptu at the Fort Bliss Noncommissioned Officers Club. He played several songs with a borrowed saxophone while we waited for our food to be served. I had supported Ron's first flight, STS-41B. Ron was a very nice guy and a good family man.

Greg Jarvis: I met Greg when he and a couple of other Hughes Aircraft employees were identified to NASA as potential payload specialists. They would operate some experiments that Hughes was proposing for flights on upcoming Shuttle missions. The missions and which Hughes employee would fly with each mission kept changing. At one time, it looked like Greg or one of his fellow Hughes employees would fly on my first Shuttle mission, which also kept changing. Greg was a nice guy and serious about doing a good job with his assigned tasks. He was easy to work with and, I think, would have been fun to fly with in space.

Christa McAuliffe: I wish I had known Christa better. She arrived in Houston fairly close to her launch date and was very busy being trained to fly in the Space Shuttle and getting her Teacher in Space Project plans pulled together. On the few occasions that we did meet, I found her to be very friendly and totally excited about the prospect of captivating students with her lessons from space, hopefully getting them excited about applying themselves in school and pursuing challenging careers. Motivating students has certainly been something that all of us in the space business have been excited about as well! I always felt somehow that Christa never fully understood the risks that she was accepting by flying in the Shuttle. I might have been wrong, but that was my feeling.

It's hard to put into words how devastating it is to lose seven people you work with and care about in an accident like this. Especially an accident that didn't have to happen.

It was a sobering time. We always knew there was the possibility for a catastrophic accident, but we had put it out of our minds and focused on our work. Very soon there was a lot of finger-pointing at the solid rocket booster joint design that was not what it should have been. Worse, there were people who knew about the design problem before the accident. All of us in the Astronaut Office were shocked, disappointed, and mad. Most, if not all of us, had never heard about this problem.

The joint design was faulty in several regards. Each solid rocket booster had four segments stacked one on top of the other. The joints where the segments met contained seals called O-rings that were designed to prevent the very high-pressure, hot gases inside from escaping through the joints during launch. Instead of maintaining a tight seal, the design actually allowed the gaps in the joints to get larger when the pressure inside the solid rocket boosters was very high during firing. Additionally, the O-ring seals were made from a synthetic rubber that became stiffer and less compliant when the ambient temperature dropped.

The Marshall Space Flight Center engineers and managers as well as the Morton-Thiokol manufacturer knew that there were design problems. They had seen undesirable erosion of the O-rings on many Shuttle flights before *Challenger*.

On the eve of the *Challenger* mission, the Marshall engineers and managers argued with Morton-Thiokol engineers and managers. They challenged the contractor to prove the solid rocket boosters *were not safe* to launch in the cold temperatures forecast for the next morning, temperatures that were outside our experience base. This approach was totally opposite to the normal process in which the government demanded that the contractors prove that their hardware *was safe* and ready to support the launch. I felt then and I still do today that those individuals in management at Marshall should have been held legally accountable for their actions.

The faulty joint design and the coldest temperatures a Shuttle had ever experienced at launch were the two major contributors to the accident. During launch, the hot gases did escape from one of the joints on the right booster. The extremely hot, high-pressure jet of gas blowing through the gap between the O-ring seal and the metal joint caused structural members of the Shuttle stack to fail, which in turn caused the entire vehicle to break up, killing the crew.

NASA had processed the video of the launch taken from the ground within days of the accident and saw the hot gas flame blowing out the side of the solid rocket booster. By May 1, enough of the right solid rocket booster had been recovered to verify the root cause of the accident, and the salvage operations were terminated.

All scheduled Shuttle missions and crew flight assignments were canceled, and all planned Shuttle operations at Vandenberg were terminated. During the two-and-a-half-year suspension of flights, the joint was redesigned, and heaters were added to the outside of the joints to keep the O-rings warmer and therefore more pliable. Several additional design modifications and assembly procedure changes were instituted to make the solid rocket boosters even safer.

In the wake of *Challenger*, several astronauts as well as engineers and trainers left the space program for a variety of reasons. Some left because of the delays; some left because their spouses had lost confidence in the program; some just left out of frustration.

I prayed and thought about what I was going to do. I talked to Karen about the situation. Eventually, Karen and I sat down with Amy and Scott at the dining room table, and we discussed it as a family. I had mixed feelings about continuing. I wanted to keep flying. I wanted to continue pushing the boundaries in space. But I had a family for whom I was responsible.

In the end, I believed that if I quit I would be letting down my friends who had lost their lives. When you look back throughout history, all great undertakings had losses, and our country could not afford to stop our space program. I had come into the program with my eyes wide open. I knew the thrill of flying in space, and I understood the dangers. I decided to do whatever I could to get us flying again as soon as possible, and my family supported me.

An incredible amount of effort was expended reviewing every procedure, every assumption, every flight rule, and all of our training requirements and methods. Our flight hardware was also scrutinized. Every design, every test and certification, every maintenance task, and every step in the preparations for launch were thoroughly reviewed. No stone was left unturned in an attempt to ferret out any other potential problems that could conceivably cause the loss of another crew and orbiter.

While the Shuttle flights were suspended, I had time to make plans for long-delayed family vacations. In the summers of 1986 and 1987, our family took two very special trips.

We had purchased a van just before my initial flight, so the first summer we loaded it up and took a forty-five-day road trip. We started at Kitty Hawk, North Carolina, toured the East Coast, and drove north through Nova Scotia until we couldn't go any farther. We traveled back west along the Saint Lawrence River and visited our families in Indiana. We had a wonderful time together.

The next summer we drove west to the Painted Desert and the Petrified Forest National Park in Arizona, the Grand Canyon, Arches National Park, Natural Bridges National Monument, Dinosaur National Monument, Grand Teton National Park, Yellowstone, Glacier National Park, and much more. After tiring ourselves out, we ended up "back home again in Indiana."

That fall I signed up for an MBA course at the University of Houston, Clear Lake. I had just started classes when I was assigned to a military mission training crew that participated in preparations for the resumption of flights. Three of the five crewmembers from the canceled Vandenberg STS-62A mission—Guy Gardner, Mike Mullane, and I—were on this crew. The other two STS-62A crewmembers, Bob Crippen and Dale Gardner, had left the Astronaut Office and were replaced on the crew with Robert Gibson and Bill Shepherd. Our military mission training crew eventually was assigned to Shuttle mission STS-27.

Because STS-27 was a classified US Department of Defense mission, all five members of the crew were in the military. That made bonding as a crew quite certain. We were very much alike, and we had a lot of fun together.

Our commander was Robert "Hoot" Gibson. Hoot is one of the best pilots with whom I have ever flown. He likes to have a good time, but he knows when to be serious. Some of the women at NASA thought he was a bit of a flirt. He would snort at the young ladies that worked on our flight checklists in an office right across the hall from ours. The women started calling him, and the rest of the crew, "swine." So we proudly became the Swine Crew, and we started wearing pink rubber pig noses to some of our training sessions. Our training team put fuzzy pink pigs on our desks!

STS-26 was the first Shuttle mission scheduled to launch after the *Challenger* accident, and our flight would follow shortly thereafter. Dick Covey, who had given *Challenger* the "Go at throttle up" call, was the pilot. The brave crew of STS-26 was constantly in the spotlight, and our second-to-fly STS-27 classified mission crew definitely was not. There was a lot of friendly jousting between our two crews.

As the STS-26 launch neared, there was a gala event in Houston, where the crewmembers were honored guests. At the end of the evening, Lee Greenwood performed his song "God Bless the USA." As he sang, fog covered the stage, and the STS-26 crew slowly rose from the orchestra pit through the fog in their tuxedos, waving to the audience. The event was shown that night on the ten o'clock news, and it was seen by several of the STS-27 crewmembers. By Monday morning, our crew was prepared.

Every Monday morning the Astronaut Office has a meeting that all astronauts and our support engineers attend. Issues are addressed, information is exchanged, and general flight schedules are presented.

Shortly after the Monday morning meeting began, I started my son's boom box, and "God Bless the USA" played loudly from the back of the room. Simultaneously with the start of the music, Hoot Gibson, Guy Gardner, and Bill Shepherd, all seated in the front row wearing white shirts and black bow ties, slowly rose from their chairs, smiling, waving, and turning to "the crowd." At the same time, Mike Mullane, who was seated behind them, recreated the "fog" by shooting a carbon dioxide fire extinguisher under the chairs of the three waving crewmembers. Everyone in the room, other than the five STS-26 crewmembers, erupted in laughter, applause, and hooting and hollering. It was one of our proudest moments!

The STS-26 crew knew they had been had! But all they could manage to say was, "Oh, yeah!" Our STS-27 crew blissfully ignored a couple of feeble attempts at counterattacks over the next couple of weeks.

The Swine Crew showed up in the Public Affairs building for our preflight press conference in our pretty pink pig noses and Lone Ranger masks. The masks referred to the fact that our mission was classified. There really wasn't anything we could say in the press conference about the mission, so we just decided to have some fun.

Hoot had phone messages from the Pentagon on his desk by the time we got back to our offices after the press conference. They didn't see the humor!

Fellow Purdue graduate Guy Gardner, our pilot, was one of NASA's tallest astronauts. Guy graduated from the US Air Force Academy and went to Purdue for his master's degree. He was in the class ahead of me at Test Pilot School and was an ASCAN classmate. Guy is a character and a lot of fun to be around. We still enjoy opportunities to get together.

Mike Mullane and I were classmates at Test Pilot School and had been friends for thirteen years by the time we flew STS-27. I've stayed in touch with him since he left NASA. Mike is a fun-seeking, gregarious guy who is always ready with a joke and a laugh. He is not athletic in a normal team-sport way, but he was a runner and still stays in great shape by climbing as many of the fourteen-thousand-foot-plus mountain peaks in the United States as he can find time to climb. Mike is smart but not professorial. He shows more than just a technical bent with his authorship of several books and his preference for classical music.

STS-27 was Bill Shepherd's first spaceflight. He went on to command the first crew to inhabit the International Space Station. Unfortunately, he couldn't tell us about most of his pre-NASA adventures as a Navy SEAL, but what we did hear made us want to hear more!

The fact that STS-27 was classified made everything we did in preparation for the flight difficult. When we traveled to receive training on our flight hardware, the travel orders were "sanitized." That meant that it took longer to process them, and only certain secretaries and other personnel with the proper security clearances could initiate them, coordinate them, or even see them.

Our training schedules were written in code. If we wanted to know what training we would be doing, we had to open a safe and use the charts inside to convert the coded schedule to plain English. All of our checklists, flight plans, and procedures were classified documents and had to be stored in safes and properly accounted for each time we completed a training session.

Our training facilities, which included gigantic simulators such as the water tank used for spacewalk training, had to be swept for electronic bugs each time we used them. The doors had to be locked and the windows covered. Everyone allowed inside during training sessions had to have the appropriate levels of security clearances. The computers used by the simulators had to be dumped and thoroughly "scrubbed" after training sessions to make sure that no classified information remained after our crew left the facility.

The scheduled time of day of our launch, the planned orbital altitude and inclination, and the planned duration of the mission were never revealed. Our actual orbital parameters were determined by amateur ground-tracking facilities as the mission was flown.

A December 1, 1988, launch attempt was postponed because of bad weather conditions at the launch site. STS-27 launched on December 2.

It was discovered before flight that one of the tires on *Atlantis* was very slowly losing pressure. During the flight, we kept the belly of the orbiter toward the sun so the sun's heat would keep the pressure in the tire high enough for landing. Things you learn in grade school science class can come in handy.

During the flight, Mike Mullane and I saw a very strange sight. We were on the flight deck, and our small attitude control thrusters were firing to maintain our orientation. The thruster firings threw out hot gases that instantly crystallized into ice particles. As the snowstorm of tiny sparkling particles flew rapidly away from the orbiter, it formed a cloud that acted like a movie screen.

We were puzzled to see a dark form projected onto this screen of ice particles. At first we didn't know what the form was, but the sun was behind us, and we realized we were seeing the shadow of our own spacecraft. It was beautiful. As the cloud flew away from the Shuttle, the shadow of the orbiter grew very quickly in size and raced away with the cloud. It was amazing to see!

Each time the thrusters fired we saw the show again. We called the rest of the crew up to the flight deck to wait for the next firings and see what we had discovered.

A normal postflight crew video summarizes the major events of the mission and shows each crewmember performing significant tasks relating to the mission's objectives. Our crew could not use the standard template to make our STS-27 postflight video. Instead, we had to film scenes of the flight that showed absolutely nothing of what we were actually do-ing during the mission. We had to be sure we did not show any classified hardware in the Shuttle's crew compartment or outside in the payload bay. Therefore, our postflight movie includes many views of normal living in space, like preparing and eating meals, shaving, and taking sponge baths.

This crew was always up for some fun! We played baseball in space and filmed it. We used a thin metal rod as the bat and peanut M&M's as balls. We pitched the "ball" to the batter, he'd hit it, and the fielder caught it—in his mouth.

Guy Gardner dressed himself up as a pirate, put a "knife" in his teeth, and swung across the mid-deck holding onto a rope like he was boarding from a pirate ship. He begged us not to show that film when we came home. So, of course, we included it in our postflight video.

Bill Shepherd, the Navy SEAL, found enough hardware on the Shuttle to build a pretty good imitation of scuba gear, complete with mask, fins, regulator, and tanks. We took pictures of him "swimming" through the mid-deck.

We all took turns at trying to "run" across the floor, up the far wall, across the ceiling, and back down the other wall to the floor. It sounds like it would be easy to do in zero-g, but it wasn't!

We had an NFL football stowed in a locker that we planned to return to NFL Commissioner Pete Rozelle at Super Bowl XXIII on January 22, 1989, in Miami, Florida. NASA had stowed it deflated. But we were a bunch of engineers! We had needles on board for rehydrating food, and we found we could use them to make the ball look like it was fully inflated.

Then we played football. We filmed it all, and it is a prominent part of the unclassified version of our postflight movie. The game started with a kickoff. "Crazy Legs" Mullane received the kickoff, was hammered immediately, and spun every which way but loose in zero-g.

We lined up against our imaginary opponents with Commander Hoot Gibson as the quarterback and me in my old high school position at center. We had the camera set up on the other side of the mid-deck, and Hoot looked right into it as he called the signals. When we huddled, we all turned our heads to look at the "formation" of the "defense." Then, of course, when we reached the line of scrimmage and got into our stance, Hoot had to wave his arms to quiet the crowd. He threw a pass, which was caught for a touchdown! The crowd went wild! The ball hung in the air for about two seconds before it was caught. Since we were flying at five miles per second, that meant the pass traveled ten miles. That has to be a record!

Finally, we kicked the extra point to win the game. We all celebrated, and being typical jocks, we came up close to the camera, waved, and said, "Hi, Mom!"

We did go to the Super Bowl after the mission, and it was a real treat. We spent some time with former astronauts Gordon Cooper,

Dick Gordon, and Pete Conrad. We met Frankie Avalon, Annette Funicello, and Christie Brinkley. Oh, yes, and we returned the football to Pete Rozzelle. It was all pretty cool.

The real postflight movie was classified. Since there were few people at the Johnson Space Center with the proper security clearances, I ended up splicing the sixteen millimeter movie of our actual classified mission activities myself. The task was made more difficult when I discovered that one of the rolls had been printed on the wrong side of the film, so the sprocket holes were on the opposite side from where they needed to be to feed correctly into the movie projector.

With no time to reorder the film before we were scheduled to go to Washington, DC, I turned over the one roll that was improperly produced and spliced it in as it was, even though the images appeared backward.

After the classified film was completed, it was hand-carried by a courier to Washington. Our crew made our first stop at the Pentagon, where we showed the movie to the Joint Chiefs of Staff. At the end of our briefing, we received a standing ovation from the Joint Chiefs—not a bad way to start one's day!

Later during our visit to Washington, we went to CIA Headquarters. There we each received a medal from the Director of the CIA Judge William Webster. Pictures were taken of each of us receiving our award, and then the awards and pictures were locked in a safe. We were told that any time we were in the DC area, we were welcome to stop by and visit our medals.

It wasn't until the end of the Cold War that each of the five of us received a package in the mail that included the medal, the citation that accompanied it, and our pictures from the award ceremony. My citation reads in part, "Colonel Ross performed in an outstanding manner, insuring that all events in the deployment and activation of a critical new national satellite system were carefully and accurately recorded and documented."

Most of the flight still remains classified even today, more than twenty-four years after the mission was completed. There has been much conjecture in the press over the years about whether Bill Shepherd and I did any EVAs in support of this mission. Certainly that is something that I can neither confirm nor deny.

I can say that STS-27 nearly ended in tragedy.

After exiting the Shuttle upon landing at Edwards Air Force Base, we did an inspection of the exterior of the orbiter and found a lot of damage. Seven hundred thermal protection tiles were damaged, and one entire tile was missing. We had a near burn-through of the orbiter's skin. We could have easily ended up the same way *Columbia* would in February of 2003. In fact, our damage on STS-27 was the worst ever recorded on any orbiter, other than *Columbia*.

Fortunately for us and for the Program, the location where our tile came off had a double thickness of metal that was just enough to protect us and prevent a burn-through of the orbiter's aluminum skin. Coming so soon after the *Challenger* accident, a burn-through could have cost more than our lives. It could have ended the Shuttle program.

My flights in space gave me some great opportunities to work with the Indianapolis Motor Speedway and the Indy "500." I flew a checkered flag for the Speedway on my first flight in late 1985. It was used during the race in 1986, and it's now in a display case at the Indianapolis Motor Speedway Hall of Fame Museum.

I met Chuck Yeager at the "500" Festival black-tie gala event in Indianapolis in May of 1986. I was the Grand Marshal of the Festival Parade, and Chuck was an honored guest. We had just been invited up onto the stage and stood in front of a large crowd. He proceeded to tell me about a pretty young women he had seen and described some of her features as looking like "popped circuit breakers." Leave it up to Chuck to compare a woman's anatomy to a part of an aircraft! I wondered if people in the crowd were curious about what he was telling me up there on the stage!

On STS-27, I flew a "500" Festival flag into space. In the spring of 1989, I returned the flag to the Festival's committee at a gathering held in a hotel at the Indianapolis International Airport. My folks were there, and Karen's folks were there, too. So was Mari Hulman George, the daughter of Tony Hulman, who had owned the Indianapolis Motor Speedway for many years and helped develop the Indy "500" into the great spectacle it is today. After Tony's death, Mari continues his tradition of announcing, "Gentlemen (and later "Lady/Ladies and Gentlemen"), start your engines," before the race.

After the ceremony at the airport, Mari invited us to join her at the track to see how they were displaying the checkered flag I had flown. So we all went to the Speedway, and she gave us a great tour of the entire museum, upstairs and down.

When the tour was over, Mari nonchalantly asked, "How would you like to go for a ride on the track in a pace car?"

I didn't have to think that one over long. We drove to Gasoline Alley where Mari introduced us to Don Bailey, the pace car driver. Don drove the pace car, Mari sat in the front, and I sat in the backseat. Dad followed us onto the track in his car with Mom and Karen's parents, Morris and "Wimpy," as his passengers. Wimpy got her nickname from the "Popeye" cartoon character who loves hamburgers.

We went around the track a couple times. Then Don stopped on the back straightaway and said, "How would you like to drive?"

How would I like to drive? Are you kidding me? Move over and give me the wheel!

I got into the driver's seat, Mari got in back, Don rode shotgun, and away we went. This 1986 Twentieth Anniversary Pontiac Trans Am really had pep. In most cars you step on it and they top out, but this car had more, and it wanted to give more. I was doing 140 miles per hour in the turns, and Don just kept smiling at me, almost laughing.

When I pulled into Gasoline Alley and we got out, I asked Don what was so funny. He took me aside and told me that Mari didn't want to ride and had never before ridden around the track with anyone but him driving.

But Mari hadn't said a word, so he figured it was all right. She must have thought if I could take the turns around the Earth at more than 17,500 miles per hour, I ought to be able to handle the banks of Indy at 140. What she didn't know was that I didn't actually fly the Shuttle. I was a mission specialist—a passenger for launch and landing. Some things are better left unsaid.

Driving at Indy was a great experience, and Dad loved it, too. He told people about driving around the track for months afterward. And every time he told the story, Mom piped in, "He was driving way too fast!"

On STS-27, with the help of Purdue University's School of Agriculture, I also flew sycamore and tulip tree seeds for Indiana 4-H. The sycamore is a favorite and ubiquitous Indiana tree, and the tulip is the state tree. After the flight, Purdue's Dr. Walt Beineke, who was known as the "Father of the Trees," and others in the Purdue Forestry Department planted the seeds and cultivated the seedlings. With the help of Purdue's Dr. Ed Frickey in the Indiana State 4-H Office, the saplings were distributed to the Purdue Extension Service offices in all ninety-two counties in Indiana. "Space trees" were planted in every county in the state, many in the county fairgrounds. Pretty neat!

In the summer of 1989, I was assigned to the screening and selection boards for the class of astronauts to be hired in 1990. We screened nearly two thousand applications. One hundred and six applicants were brought to Houston for the week-long series of physical exams and the all-important Astronaut Selection Board interview.

I entered the process determined to pick the very best people, but I quickly decided that I couldn't quantify "the very best." Instead, I decided that I would attempt to only hire individuals with whom my astronaut friends and I would be comfortable flying. That approach was much more realistic, and I am glad I used it as my guide.

We hired twenty-three astronauts in the 1990 class, and I ended up flying in space with nine of them. I was always very proud that we selected Eileen Collins to be the first female pilot astronaut in the history of the US space program. Eileen went on to an illustrious career as an astronaut and the first female Shuttle commander; she is an outstanding role model for my daughter and other young women everywhere.

In the fall of 1989, our daughter, Amy, left for Purdue. That's when fate stepped in and we had a blessed arrival at our home, a new member of the family.

It was a parrot named Sunny. I'm not the only flyer in our family.

I had always thought it would be neat to have a parrot, but Karen didn't agree. Okay. Dead subject. I didn't bring it up again. However, one day in January, Karen called me and said we were going to look at a parrot. Sunny was about seven months old. The family who owned him had a toddler who was allergic to feather dust, so they needed to find a new place for Sunny. We brought him home.

Since we got Sunny the same year Amy left for college, she's always called him her "replacement unit." He's turned out to be quite a character, and he certainly thinks he runs the house. And he might be right.

He talks quite a bit, including telling us to get off the telephone by saying "Good-bye" when he thinks we've talked too long. He sings. Karen taught him most of the songs, but we all helped. He's really very good. The first song he learned was "Take Me Out To the Ballgame." He certainly heard it often when I watched Cubs games on TV. At Wrigley Field, they always sing it in the middle of the seventh inning. He loves to sing duets. We sing "Jesus Loves Me," "Row, Row, Row Your Boat," "My Bonnie Lies Over the Ocean," and "Jingle Bells." Sunny learned "Old MacDonald Had a Farm" in one day. He even sings the national anthem, and he can carry the tune better than I can.

My mother once let Sunny out of his cage only to discover that our cat was still in the house. "Kitty" came downstairs into the family room. Sunny flew off his cage, landed on the floor close to the cat, and said, "Here, kitty, kitty, kitty . . ." Mom almost fell over, and Kitty turned and headed back upstairs.

When we let Sunny out of his cage, he flies and walks or struts around the house. We keep an old piano in the dining room just for him, and he walks up and down the keyboard playing music. He's not bad! Amazon parrots can live to be sixty years old or older, so the kids argue about who inherits Sunny—or who might be forced to inherit him.

While we were getting to know Sunny, I was preparing for my third flight, STS-37. The main objective of STS-37 was the deployment of the Compton Gamma Ray Observatory.

NASA built four satellites that were collectively titled the "Great Observatories," and three were launched on the Shuttle. The first was the Hubble Space Telescope, which observes the universe in the visible and near-ultraviolet frequency spectrums. The second was the Compton Gamma Ray Observatory, which studied gamma ray and hard X-ray sources in the universe. Third in the series was the Chandra X-ray Observatory, which studies soft X-ray emissions. The fourth telescope, the Spitzer, was launched aboard an unmanned rocket in 2003 and studies the infrared spectrum.

The Gamma Ray Observatory was high priority. At thirty-five thousand pounds, it was at that time the heaviest satellite to be launched into low Earth orbit by the Shuttle. The first time I saw it I couldn't believe its massive structure. It didn't look like a delicate satellite. Most of the time when viewing a satellite at close range, it looks so fragile that one is afraid to even breathe on it; however, the Gamma Ray Observatory looked more like a diesel locomotive. Everything on it was bulky, thick, and heavy.

Gamma rays are an invisible, high-energy form of radiation that can't penetrate the Earth's atmosphere. By studying gamma rays, scientists are able to understand more about the Milky Way, pulsars, quasars, black holes, and the origin of the universe. This was pretty exciting stuff for astronomers and a stargazer like me.

STS-37 launched on April 5, 1991. Our commander was Steve Nagel and the pilot was Ken Cameron. Jay Apt, Linda Godwin, and I were the three mission specialists. Steve was also flying his third mission. The rest of the crewmembers were on their first flights. Linda and Jay were hired in the 1985 class of astronauts. Their class was the third class with no nickname. Since the 1984 class ahead of them, which included Ken, was nicknamed the Maggots, the 1985 class tried calling themselves "The Class with Class," but it was never "officially" recognized. It is against the rules to give yourself your nickname!

Steve is a really nice guy, but I normally won't admit it. He has one big problem. He went to the University of Illinois, a strong Big Ten rival of Purdue. Steve and I have had a lot of fun over the years with our ongoing Purdue-Illinois banter.

NASA had not performed a spacewalk since my STS-61B flight in 1985—almost six years earlier. We were preparing to build a space station and therefore needed more EVA-experienced astronauts to help develop and perform EVAs.

I told Steve that of course I would like to do the mission's planned EVA, but since we were trying to get more EVA-experienced crewmembers, I would understand if Steve picked Jay and Linda to do the EVA. I also told him that I thought the likelihood of having to do an unplanned EVA to work on the Gamma Ray Observatory was extremely low. Steve thought about it for quite a while before reaching a decision. "You know, I understand what you're saying," he said. "I think it's a good thing to consider, but I'd really look stupid if we had to do an unplanned EVA on the Observatory and you weren't one of the guys outside." I was going to spacewalk again!

We had several slips in our launch date as we prepared for STS-37. Launch delays were always something with which flight crews had to deal. When we were still months away from a planned launch date, the delays were annoying, but when the delays happened only weeks or days before a launch, they were painful.

The pain was felt more by families and friends of the crewmembers than the crewmembers themselves. Family members and friends had to coordinate vacation time, arrange for children to be out of school, and make lodging and airline reservations well in advance of a launch date. When we slipped, family and friends had to decide if they could still travel to see the launch and then rearrange everything. Frequently, we had launch delays after the crew and all of our guests were at the Cape. Most of these delays were caused by bad weather either at the Cape or the contingency transatlantic landing sites, but sometimes it was a last-minute Shuttle hardware problem that caused a delay.

Two very unexpected events occurred during the training flow for STS-37 as a result of delays in our scheduled launch date. One was an incredible opportunity that came straight out of the blue, and the other ended up with me spending time doing something I really didn't want to do.

The incredible opportunity was actually the result of years of quiet space diplomacy. In July of 1982, former astronaut Rusty Schweickart met informally with cosmonauts Alexei Leonov, Vitaly Sevastyanov, Georgi Grechko, and key Soviet officials to explore establishing an organization of those who had flown in space. Additional meetings in 1983 and 1984 led to the establishment of the Association of Space Explorers (ASE) in 1985. ASE was founded to provide a professional forum for individuals who had flown in space to support science and exploration for the benefit of all, to promote education in engineering and science, to raise the awareness of our planet's environment, and to encourage international cooperation in the human exploration of space. Soviet and US chapters were established.

Late in 1989, the Soviet ASE chapter invited the US chapter to send four astronauts to Moscow for a week-long visit to the Soviet space facilities. Former astronaut and then the Johnson Space Center Deputy Director Paul Weitz, Chief of the Astronaut Office Dan Brandenstein, experienced Shuttle commander Ron Grabe, and I were selected for the trip. Because STS-37 had been delayed, I was available. This may be the only time I was ever happy a launch slipped.

We arrived in Moscow on February 9, 1990. The Communists still ruled even though several cracks were showing in the Soviet Union. Moscow was gray, with low clouds, and cold. After living in Texas for more than ten years, my blood had definitely thinned.

While packing for the trip, I noticed that I no longer had a good, warm winter hat. I decided I would wait until I arrived in Moscow to buy one. Where would be a better place to buy a winter hat? Certainly not Houston.

Former cosmonauts General Vladimir Kovalyonok, Oleg Makarov, and Alexei Leonov were the primary hosts during our visit. After depositing our belongings in the hotel, we gathered in the lobby to depart for General Leonov's art studio and home for dinner. I explained my plan to buy a hat in Moscow to Alexei and asked where I could get one like his.

"No problem," he said.

We boarded several minivan-like vehicles and headed downtown. We finally pulled up in front of a large, gray, heavy-looking building in central Moscow. People clad in military attire were entering and exiting. Our vehicles parked in a haphazard manner directly in front of the doors, and we all unloaded and went in. Once inside, it was apparent that this was not your normal Macy's. We were in the Red Army's clothing store! The two generals led with the rest of us parading along behind. We went upstairs and marched up to a glass showcase with hats inside. Not understanding Russian, I still got the gist of the demand from the generals to the young lady behind the counter: "Give all of our American friends hats!"

It was also obvious that the young lady was apprehensive dealing with the strange demand from the two generals. She was not particularly eager to do what they asked, but she was also worried about not honoring their demands. She wanted to speak with her manager first. The generals did not want to wait. After more "discussion" and waving of arms, the rest of the American delegation and I all had new, very warm, Soviet military hats. Great, I had only been in the USSR a matter of a few hours and had already nearly caused an international incident—over a hat.

Our tours of the Soviet Union's space facilities during the next week were overwhelming in the level of access and detail we received. The day after arrival we boarded a Soviet Air Force passenger jet and flew from Moscow to Baikonur in Kazakhstan. My travel companion during the flight was General Gherman Titov, the second person to orbit the Earth and the fourth person ever to fly in space.

Once at Baikonur, our group toured the launch pad that would be used the next day to send Anatoly Solovyev and Aleksandr Balandin aboard Soyuz TM-9 to the Soviet Mir space station. The Soyuz spacecraft and its launch vehicle had just arrived at the pad the previous day, only two days before the launch. While there we were shown a small monument not far from the pad that was made from material left over from the construction of the *Sputnik 1* spacecraft. The next day's launch was to be from the same pad where that first satellite was launched on October 4, 1957.

We visited the facilities where several Burans, the Soviet Union's near carbon copy of the Shuttle, sat collecting dust. Buran looked almost exactly like our Shuttle, but there were some differences. One of the main differences was that Buran only flew in space once, made two orbits around Earth, and had no people on board. That flight was in 1988 when the Soviets were having tremendous financial problems.

The Buran that flew was destroyed in 2002 when the hangar in which it was stored collapsed.

We toured the heavy-lift Proton rocket processing and launch facilities. We also met briefly with the TM-9 flight crew in front of their quarantine facility, which serves the same purpose that our Astronaut Crew Quarters at the Cape did prior to Shuttle launches. We walked along a long path lined with trees planted by previous Soviet crews while waiting for their launches, a tradition still followed today.

The next morning we were up early to watch Solovyev and Balandin suit up and dutifully follow formal protocol with senior officials before boarding a bus and heading to the pad. While the crew was getting settled into the Soyuz capsule, our guides took us on a tour of two small cottages where Yuri Gagarin and Sergei Korolev, the mysterious Soviet "Chief Designer," slept the night before Gagarin's momentous flight. In the early days of the Space Race, the Soviets would not even let the world know who was leading their program. He was simply known as the "Chief Designer." We also toured a very nice museum nearby, and then as the launch time approached, we were taken to an outdoor viewing area relatively close to the launch pad. It was extremely cold, and there was a low ice fog over most of the launch complex. But at the appointed time, the Soyuz rocket ignited and smartly climbed into the blue sky, leaving a white streamer of smoke in its wake.

We returned to Moscow later that day, and the following day we were taken to Star City, the military facility to the north and east of Moscow where the cosmonauts train for their missions. We toured their offices, training classrooms, medical facilities, centrifuge, and gymnasium. Then we went to the large facility where a full-scale trainer of the Mir space station and the Soyuz trainers where located. We were given a surprisingly detailed tour of both facilities and were allowed to ask many questions. The US pilot astronauts were invited to fly the Soyuz simulator, so I asked if I could possibly get into one of the Soviet spacewalking suits. It was arranged, and I was thrilled with the opportunity to gain firsthand knowledge of their suit. I completed a suited evaluation before lunch and found their version to have advantages and disadvantages compared to ours. Their suit was easier to don and doff, but it was stiffer and harder to work in. The mobility of the suit's arms and the flexibility of the gloves were notably worse than ours.

After lunch I was given a nice surprise. During earlier discussions, the Russian spacewalking experts learned that I had done developmental work with our Manned Maneuvering Unit (MMU) rocket backpack. Over the lunch hour they arranged for me to have an opportunity to fly the simulator for their version of the MMU, which they called their "space bicycle."

To fly the simulator I again donned one of their space suits and was backed into a device that simulated their "space bicycle." I was positioned in front of a video screen. The views on the screen changed in response to the inputs I made into the "space bicycle" controls. I found their version of the MMU harder to fly than ours, primarily because it always had a tether attaching it to a structure. That meant I had to manage the tether as well as fly the "bicycle." I think with additional training, I would have been as comfortable flying their device as I was ours. I felt like a kid in a candy shop!

Later that day we were invited to a party to celebrate the successful launch of the Soyuz. The guests of honor were Anatoly's wife, Natalya, and Aleksandr's wife, Lydia. The party started in the late afternoon with lots of vodka and toasting and no food. I had never drunk vodka before this trip, but in order to not offend our hosts, I sipped. And every time I sipped, they refilled. Somewhere along the way my American colleagues informed me that, as the junior member of our group, I had been elected by them to sit at the head table and be responsible for giving a toast.

We had already been drinking and toasting for a couple of hours by the time the guests of honor arrived, and it was still some time before everyone started to move toward the tables to find their seats for dinner. I was seated at the head table as directed, close to General Alexei Leonov, who was the emcee for the evening. Alexei is a force of nature. He commands respect in any group and has an exuberant, fun-loving personality. On March 18, 1965, General Leonov had become the first human to do a spacewalk; I felt very honored to be with him.

Shortly after the food was served, I realized that if I was going to be able to stand and make a proper toast, I had better do it soon. At the same instant that I started to stand up to propose the toast, General Leonov also stood up. Fortunately for me, General Leonov understands English quite well and also has a great sense of humor.

I turned to him and said, "General, you need to sit down. I want to make a toast." I paused, and then continued, "And if you don't let me do it now, I might not be able to stand up or speak later."

Alexei laughed, smiled one of his bright smiles, and explained to the other guests in Russian what I had said. After the laughing stopped, I gave what I was later told by Americans and Russians was a very nice toast!

The next day was the scheduled docking of the Soyuz TM-9 with the Mir space station. We were taken to a suburb of Moscow, which was then called Kaliningrad, where the Russian Mission Control Center is located. Their Mission Control Center is part of the sprawling complex of the Energia Design Bureau. Sergei Korolev, the secret "Chief Designer," founded the Bureau. We sat in the balcony of the large Mission Control Room and watched the big display boards, which are much like the displays in our Mission Control Center. Through interpreters, our hosts explained the displays and how the rendezvous of the Soyuz with the Mir station was being performed.

After the successful docking, our hosts took us on a field trip. We traveled north of Moscow to the town of Zagorsk, now known as Sergiyev Posad, and the monastery Trinity Lavra of St. Sergius, the spiritual center of the Russian Orthodox Church. After lunch with an archbishop, we were taken on a tour through the incredibly beautiful churches and buildings, most of them topped with gold and blue onion domes.

The final day of our visit was spent as tourists seeing the Kremlin and its Treasury, Red Square, and the surrounding area in downtown Moscow. That evening we went to the Great Moscow State Circus. The Circus was a real delight with all types of very talented performers. I was informed that it was a "requirement" to have ice cream at the circus—it was delicious!

I left Moscow with a much clearer and better impression of the Soviets and their space program. Certainly they did some things differently from us, but they had good reasons for doing so, and the bottom line was—their hardware worked.

I also felt sorry for the Soviet citizens because they were burdened with such a miserable lifestyle due to the Communist Party rule. Very few people smiled. Everyone dressed in shades of gray, black, or brown. All of the buildings were of the same hues. There were no bright colors in their lives.

Driving around Moscow, we saw long lines of people standing outside at night in the bitter cold waiting to get into stores whose windows were papered over. When we asked our interpreters what the people were waiting for, the interpreters told us that the people did not know what was in the stores. Even if they didn't need whatever they found inside, they would buy it and then see if they could trade for something they did need. The cosmonauts drove vehicles that were in poor condition. We had a flat tire on one vehicle the first night we were there. Another vehicle had a bad wiper that would go across the windshield but not back. The driver had to pull it back with a string tied to the blade.

Visiting the Soviet Union was the incredible opportunity.

Next came what I didn't want to do.

Ken Cameron, a Marine, was the rookie pilot on our STS-37 crew. Ken was an amateur radio, or ham radio, operator. He had a passion to include a ham radio experiment on our flight. He accomplished his first goal; the Shuttle Amateur Radio Experiment (SAREX) was added to the mission. His second goal, which was not announced to the crew until after the first goal was achieved, was to have all of the crewmembers become ham radio operators and speak on the radio while on orbit.

Fellow spacewalker Jay Apt, who was actually pretty gung ho about the idea, studied and passed the examinations required to be a licensed ham radio operator. Commander Steve Nagel and Mission Specialist Linda Godwin were no more than lukewarm to the idea. But after one of our early schedule slips, Ken was able to convince them to study, and they passed the tests. I was the lone holdout on the crew, preventing

Ken from achieving his second goal. I felt that NASA had paid good money for all of the radios installed in the Shuttle, and they worked really well. I didn't need another, less capable one.

Marines don't take well to not achieving a goal and can be extremely persistent in completing their mission. Ken would place the study materials for the ham radio license on my desk, right in the middle where I couldn't miss them. I would place them on the corner of my desk and ignore them. Each day they were back in the middle of my desk, so I started hiding them somewhere in our office or even putting them in the trash can. Each morning they would show up back on my desk, in the middle. In addition to the materials being always around, Ken would routinely give me verbal reminders, always with a coaxing smile. The reminders came during training classes, while flying in T-38s, or even in the middle of a simulator session with malfunction alarms going off all around us.

Finally, when the flight date drew near and I thought the likelihood of our launch date slipping yet one more time was pretty remote, I broke down in the face of yet another full-frontal attack from Ken one morning in the office. The whole crew was there. Ken started in on his sales pitch, and I said, "Ken, if we slip again, I will get my ham license." Ken's eyes lit up like a Christmas tree, and he almost jumped over his desk to hug me. At the same time he was getting confirmation from the rest of the crew that they had indeed heard me make my commitment so he would have witnesses, if needed. Of course, shortly thereafter our launch date slipped, and I very begrudgingly took the study materials and prepared to take my ham radio license tests.

The license exam had two parts. The technical written exam was not very difficult, but the second part required me to be able to receive and translate a message transmitted in Morse code at a specified speed. To prepare for the second part of the test, I had to memorize the Morse code, which represents each of the letters in the alphabet with dots and dashes, and then listen to a bunch of short and long beeps coming over a headset. This part was not fun for me, and I didn't like spending the valuable time before launch doing it. But I had given my word.

I passed the tests and got my license and my very own ham radio call sign. My call sign was N5SCW, and I never used it on the ground. Once we got on orbit, Ken couldn't wait to set up his "ham shack" in and around his pilot's seat on *Atlantis*'s flight deck. There were cables, headsets, electronics, and antennas everywhere. It looked like a snare just waiting to entangle the next "Bigfoot" who ambled by.

Whenever he was free of other duties, Ken talked on the radio to another ham on one of Earth's continents. Jay also conducted several conversations on the amateur radio. Linda and Steve tried their hands at it as well, and even seemed to enjoy it. Will wonders never cease! I avoided the radio like the plague.

We were quickly approaching the end of the mission, and Ken soon faced tearing down and stowing his ham radio equipment. He finally came to me with great big, puppy-dog eyes and asked if I would make just one call on the radio before he had to put it away. How can you refuse a Marine who looks so forlorn? I put on the headset, made a call, and waited for someone to respond. I briefly talked to someone on one of the continents. I don't remember who or what continent, but Ken was elated. So was I, because since his goal had been accomplished I thought maybe I could have some peace and quiet! I am probably the only ham radio operator in the entire history of the world who has never made a ham radio contact on the surface of the Earth, but has made one, and only one, from Earth orbit.

Years later at a crew party, Ken, with a big grin, revealed to all of us that during the time I was stalling, the requirements for the amateur radio license had changed. I could have received my ham license without taking the Morse code test. Now he tells me!

The Compton Gamma Ray Observatory was scheduled for deployment on the third day of our mission. Jay and I were prepared to do a spacewalk if there was a problem, but, of course, we all hoped everything would be fine.

When it came time, Linda positioned herself at the robotics control station on the aft flight deck, looked through the aft windows, and used the robotic arm to lift the Observatory out of the payload bay. The task went perfectly.

I was floating in the pilot's seat. Steve was next to me in his commander's seat. Jay, Linda, and Ken were behind us on the aft flight deck. Each solar array opened, one at a time, just as planned. These were the events that we all thought most likely, if any, to have problems.

Jay piped up, "Well, it's all downhill from here."

All of the rest of us admonished Jay for his premature comment. The next and last major activity to be performed prior to releasing the satellite from the robotic arm was the deployment of the long Ku-band antenna boom. And sure enough, the antenna boom on the satellite wouldn't come out. It was stuck.

We had procedures to try to free it. None of them worked. Frankly, none of us expected them to work. While we were making our second attempt to free the antenna, I took off my wedding ring and looked at Steve. "I'm going downstairs to get ready for an EVA," I said. "I think that's the right thing to do," he agreed.

At the same instant in West Lafayette, Indiana, Amy was watching NASA TV in her residence hall at Purdue. "He's going to do a spacewalk," she said.

In less than a minute the orders came from Houston to Steve: "Why don't you send Jerry and Jay downstairs to get ready for an EVA."

This was going to be my third spacewalk and my first unplanned EVA. I was a little nervous. If we couldn't fix the $670 million Observatory, we would have to deploy a very large and very expensive piece of space junk.

Once Jay and I were outside the airlock, I asked Linda to move the Observatory down and closer to the starboard (right) side of the payload bay. Linda operated the arm to bring the satellite into position. Then I looked for a path so I could move to where the antenna boom was mounted.

The satellite's large propellant tanks were right in front of me, and I knew I had to be very careful not to rupture them or to do any damage to the satellite while moving around. I carefully made my way to the back side of the Observatory and approached the antenna boom. I inspected the boom and its latch as thoroughly as I could and reported to the rest of the crew and Mission Control that there was no readily apparent reason for the boom's failure to deploy. I also communicated that I thought I was in a good position to be able to apply a significant force onto the boom and still control my body position to prevent collateral damage. After a momentary loss of communications with the ground, Mission Control came back with their approval for an attempt to free the boom by pushing on it.

I got into position. Steadying myself with my left hand on the Observatory structure, I reached out with my right hand and gave the boom two approximately forty-pound pushes. The boom didn't move. I pushed a third and fourth time, and it moved a little. Finally, with the fifth and sixth pushes, the boom swung free! Then Jay and I fully deployed and locked the antenna boom in its on-orbit position.

I was very relieved and privately exuberant. The rest of the crew was excited, and the ground team was elated. A good old country boy's encouragement had done it! And it only took seventeen minutes. I earned my paycheck that day. We also demonstrated the importance of having humans on spaceflights. Only humans can perform contingency procedures when things don't go as planned.

After we fixed the satellite and before it was released, I came up to the large windows that looked out from the aft flight deck into the payload bay. Steve Nagel took three pictures of me with a great big smile on my face looking in one of the windows. The first and third pictures did not come out at all, but the second one was great. It was used in all kinds of articles and magazines, and I received copies of a travel magazine from friends all over the country that used "The Picture" on its cover. For several years that photo of me from STS-37 welcomed travelers to the Space Coast as they arrived at the Orlando International Airport.

Jay and I were supposed to be fully into the airlock when the rest of the crew ultimately released the Observatory, but I think the only parts of us that were in the airlock were the tips of our boots. We stayed outside in the payload bay to watch. We were flying over northern Africa. The orbiter's open payload bay was facing Earth, so we could see the Observatory underneath us and beyond that the ground. It was a marvelous sight.

The next day on the planned EVA, Jay and I tested a piece of prototype equipment being considered for the International Space Station. I had conceived of the device called the Crew and Equipment Translation Aid (CETA) cart several years earlier and had won its inclusion in the design of the ISS. The cart would run up and down a track on the truss of the ISS and make it easier and faster for spacewalkers to move themselves and their tools and equipment from place to place.

Jay and I had led a campaign for NASA to start doing more spacewalks in preparation for assembly of the ISS. We had worked with the Space Shuttle and Space Station programs to select STS-37 as the right flight to start adding EVAs and helped to select the CETA cart as the hardware to be flown. Shortly after those two decisions were made, Jay and I were assigned to the STS-37 mission. We didn't have any idea that we even were being considered for the flight, but we were thrilled.

We worked with the design engineers to develop three different types of carts and the deployable track on which the carts would travel. One cart was manually powered. A second cart was mechanically powered and looked and functioned much like a handcar on a railroad track. The third cart was electrically powered with a hand-cranked generator that provided electric power to a small motor.

We developed the hardware in the water tank and then used the same facility to train for the EVA. The manual cart and the mechanical cart both worked in the water tank. Obviously, we couldn't put a fully functional electric cart in the water. The electric cart operation could only be simulated, and the scuba divers supporting our training actually pulled the electric cart up and down the track in response to verbal requests from the spacewalkers.

In space, we set up a track along the side of the orbiter's payload bay, and we ran the tests. The simple manual cart worked perfectly. The mechanical cart was okay but awkward to use. But the electric cart gave me fits. I couldn't get it to move. I cranked and I cranked the generator, but nothing happened. But I kept giving it the old college try.

I cranked the electric cart so hard that the rest of the crew told me later the whole orbiter was shaking. I completely wore myself out. Finally, I stopped for a break before one last attempt. I looked down at the cart.

The parking brake, which we never used in the water tank, was on.

Now, if I had thought faster, I would have radioed, "Well, the parking brake works!" Instead, I told them what I'd done. I released the parking brake, and the cart rolled nicely down the track. And I got a really good dose of teasing.

Later, near the end of the spacewalk, I was in a foot restraint on the end of the robotic arm high above the payload bay.

It was April 8, 1991. The rest of the crew was working with Jay, and I had a few moments to myself. It was dark, and I turned off my helmet lights. I leaned back and stared into the blackness of space. Suddenly, I had a strong sensation that I was at one with the universe and that I was doing exactly what God had designed me to do. I was using my brain and my hands to work in space fixing spacecraft, building structures, and testing ISS equipment. I had an overwhelming sense of fulfillment and affirmation about what I was doing. For me, an engineer, to have an experience like that—it was just incredible.

When we got back inside after the EVA, Jay took his gloves off and noticed a large bloody spot on one of his hands. I thought he had broken a blister. It turned out a metal part of his glove had punctured the bladder and was rubbing against his skin. Fortunately, the bladder maintained a seal so he didn't have to come back inside early with a leaking suit. We were lucky on that one.

After the mission, Jay and I entered a room filled with Goddard Space Flight Center engineers and managers responsible for the Gamma Ray Observatory to debrief them. I had obtained a pip pin with a red "Remove Before Flight" banner on it. It was the same type of pip pin and banner that is frequently used on a spacecraft to secure critical systems until the pins are removed just before the spacecraft is launched.

We acted as if I had found the pin attached to the antenna boom, and their failure to remove it was the reason the boom had not properly deployed. When I tossed the pin on the table and asked them if they wanted it back, the looks on the managers' and engineers' faces were priceless; I thought some of them were going to pass out. The effect was so dramatic that we had to immediately break into big smiles and assure them that we were was just kidding!

They all saw the humor. Eventually.

President George H. W. Bush invited the crew of STS-37 and our spouses to the White House. We met President Bush in the Oval Office. He was a most congenial host and was very interested in our mission and supportive of space exploration in general.

Walking into the Oval Office is a wonderful experience. It's something you dream about. It's not as big as I expected—it's a very comfortable size. When we arrived, President Bush was standing to greet us. He shook our hands, thanked us for stopping by, and asked about our flight. He told us about the history of his desk and the Oval Office. He then proceeded to take us on a tour that included the Rose Garden and the pit where he played horseshoes. He was a delightful person.

The White House Chief Usher came to take us up to the private family living quarters, including the Lincoln bedroom. President Bush apologized that his wife, Barbara, was not home to lead the tour. She had just taken his mother to the family home in Kennebunkport, Maine.

Just before we left the Oval Office, President Bush handed each of the other crewmembers a small box containing a gift. But since I was across the room and behind a couch at that moment, he casually tossed a box to me across the middle of the Oval Office.

I was surprised. I didn't expect the President of the United States to toss anything, but at the same time instincts kicked in and I caught the box. President Bush, a former first baseman at Yale, flashed an approving smile.

The box contained a tie clasp that displayed the Presidential Seal. I enjoy having the tie clasp and cherish remembering the day I played catch with President Bush in the Oval Office.

My journeys in space had been extraordinary. I had flown on three missions and had completed four spacewalks. And two more flights were right on the horizon.

What I didn't know was that it would be seven years before this spacewalker would walk in space again.

SEVEN

Two more boarding passes

O n the drive from Astronaut Crew Quarters to the launch pad, the
Astrovan always stopped at the Launch Control Center, where
management personnel disembarked. I know some of them would have
liked to go with the crew.

Sometimes as the Astrovan slowed to a stop, one of the managers paused, took a look around at the crew, and perfunctorily said, "I'll
need to collect your Shuttle boarding passes now before I can allow
you to proceed to the pad."

Experienced crewmembers reached into a pocket on their launch
suits and pulled out identical, official-looking Shuttle boarding passes.
The rookies instantly panicked and frantically searched through the
pockets on their suits to find a pass they had never heard of or seen before.

The rest of us watched with glee and held back laughter as long as
we could. It usually wasn't long. When the rookies finally picked up on

the smiles and growing laughter, they realized they'd been had! It was a great "gotcha moment," and it helped to break the tension of the day.

The laugh at the expense of the rookies was all in fun, but in truth, getting a "boarding pass" for a ride on the Shuttle was not an easy process. But to my surprise, in the spring of 1991 prior to STS-37, once again I was assigned to another flight before launching on my current one.

Dan Brandenstein, Chief of the Astronaut Office and the prayer leader for STS-27, called me while I was in quarantine before the Gamma Ray Observatory flight. Our crew had just watched on television the FBI assault on David Koresh and the Branch Davidians in Waco, Texas, and I was trying to deal with images of fire and people dying when the call came. My mind wasn't on spaceflight.

Dan wanted me to fly as the payload commander on a German-sponsored D-2 Spacelab mission. A payload commander led the experiment team and had responsibilities for the payload similar to the Shuttle commander's responsibilities for the overall success of the mission. I told Dan I thought the payload commander for the D-2 mission should be an astronaut with a science background instead of an engineer like me; however, he assured me I was the right guy for the job. So I told him I'd do it.

While our STS-37 crew was still completing postflight debriefings, my assignment as payload commander for the STS-55/D-2 mission was announced. STS-55 was NASA's flight number, and D-2 was the designation for Germany's (Deutschland's) second Spacelab mission.

The Spacelab was a European Space Agency contribution to the Shuttle program. It was a science laboratory module designed to fit into the payload bay of the Shuttle. When it flew, Spacelab was connected to the Shuttle's crew compartment mid-deck by a pressurized tunnel so crewmembers could easily move between the two.

Spacelab provided the electric power, computer controls, data collection, and other types of laboratory support to individual experiments. It sent experiment data and TV views of what we were doing in the Spacelab to the Shuttle, which relayed the signals to the ground. Spacelab had standard-sized spaces where experiment racks could be installed. It also provided considerable space for the stowage of experiment supplies and equipment.

In addition to the Mission Control at Johnson Space Center, payload mission control centers were located in Alabama at the Marshall Space Flight Center and in Germany and Japan. The Spacelab crew of four astronauts normally worked in two shifts around the clock and routinely communicated directly with the experiment specialists located in the payload control centers.

Most of the experiments utilized the unique environment of zero-gravity to aid in their research. All types of experiments could be conducted. On D-2, we performed experiments in fluid and solid-state physics, chemistry, crystal growth, and metallurgy. We also investigated the effects of zero-g on seed germination and plant growth, the behavior of small animals like fish in zero-g, and how the human body changes over time in the weightless environment.

Just a month after STS-37 landed, I was headed to Germany to meet the team preparing for the Spacelab mission. I received a series of briefings over a ten-day period about each of over one hundred experiments planned for the mission and the training associated with them. It became readily apparent that the number of experiments planned would not fit into the crew time available during the mission and that the training would have to be significantly reduced to fit into the one and a half years available before launch.

The main reason we were overbooked with experiments soon became clear. The German government bought the flight and planned their experiments. Then, to reduce their costs, they sold some mission time and payload space to the European Space Agency and to NASA. Those two agencies each added experiments, but the Germans did not reduce theirs.

This proved to be a chronic source of problems throughout the preparations for flight. The German mission manager was between a rock and a hard place, trying to satisfy his management as well as the German, European Space Agency, and NASA experiment teams. Consequently, the problems trickled down to the crew—and therefore to me, the Payload Commander.

The rest of the Spacelab payload crew—NASA Mission Specialist Dr. Bernard Harris, Jr., and the German Payload Specialists Hans Schlegel and Professor Ulrich Walter—were rookies. My bottom line goal for the mission was to achieve a maximum science return and at the same time to keep the mission from being so overbooked that we would fail to accomplish all assigned tasks, no matter how hard we tried.

For the next year and a half, Bernard and I traveled to Germany every other month for training. We trained on the theory and the practical aspects of each of the experiments. In addition, we reviewed and rewrote the procedural checklists and malfunction checklists. We did at least two revisions on most of these documents. The German payload specialists and Bernard and I spent the other half of our time in Houston training on Spacelab and Shuttle systems.

Bernard and I found an apartment on the third floor of a house high above the Rhine River near Königswinter. The house bordered a vineyard, and we routinely walked along the vineyard to the little town of Rhöndorf. Rhöndorf is on the east side of the Rhine just south of Bonn and a mile or two north of where the famous World War II Ludendorff Bridge at Remagen stood. The bridge was captured by US troops in March of 1945, just before the end of the war. It was the only remaining bridge over the Rhine River not destroyed by the Germans, giving Allied forces easy passage into the German heartland. Bernard and I frequented a small restaurant at the Hotel Zinn-Rebengarten in Rhöndorf and especially enjoyed their schnitzel, pommes frites, and a beer or a glass of halbtrocken white wine. The owners became our good friends, and after the mission we had a big party at their restaurant.

Finally, launch day arrived. We were exhausted but excited and anxious to go. Many guests and VIPs had come from Germany to see the launch and were at the viewing sites. I was seated in the mid-deck with Hans and Ulrich. Seconds before we were to lift off, the Shuttle's main rocket engines started, then stopped. We'd had a launch pad abort. The Shuttle's computers had sensed a problem in one of the three main engines and halted the countdown at three seconds. I could feel the stack swaying. I unstrapped from my seat and moved to the side hatch. I wanted to be ready to quickly exit the orbiter if the Launch Director ordered us to evacuate. The order never came, and we waited until the closeout crew came back to the pad and assisted our exit.

After the pad abort, the payload crew returned to Germany for more training and medical tests, which was really the last thing I wanted to do. When we got back to Houston, I arranged for the whole crew to go see the movie *Groundhog Day*, in which actor Bill Murray (a fellow Cubs fan) relives the same day again and again. I thought it was appropriate considering that shortly we would go back into quarantine and make another trip to the Kennedy Space Center for launch.

A little more than a month after the launch abort, we finally blasted off on April 26, 1993, but with many fewer people watching.

The flight crew included Commander Steve Nagel, who had also been my commander on STS-37, Pilot Tom Henricks, and Charlie Precourt as mission specialist two. They got us to orbit, brought us home, and kept the orbiter operating while the payload team worked in two, two-man, twelve-hour shifts. Steve, Tom, and Charlie also operated some experiments mounted out in the payload bay and participated in some studies of the human body in zero-g.

Steve knew how much I loved to look out the orbiter's windows at the Earth. He also knew that I was going to be extremely busy and that there were no windows in the Spacelab. To tease me, before the flight he arranged for a copy of "The Picture" he had taken of me looking in one of the orbiter's aft windows on STS-37 to be mounted behind a cover over that same window in *Columbia*. Steve knew I would be the one removing that cover in preparation for opening the payload bay doors shortly after arriving on orbit.

His plan worked perfectly. Unsuspecting, I removed the cover and came face-to-face with myself looking in at me. I burst out laughing. He got me! Later Steve brought the picture back into the Spacelab and put it on the end wall of the lab. He said I could pretend I had a window. Isn't that what friends are for?

The first time I went to Germany I learned that almost one-half of the experiments we would be doing involved collecting blood, saliva, and urine from the payload crew. All of these samples would be preserved in a freezer in the mid-deck of the Shuttle for analysis later, on the ground. I knew the Shuttle freezers were notoriously unreliable, so I encouraged the German program manager to add a second one.

Even though storage was already very tight and another freezer would occupy the space of two mid-deck lockers, he reluctantly agreed. It was a good thing. The primary freezer was operating at liftoff, but it had already failed by the time we got to orbit. We activated the second freezer and pampered it throughout the mission. If it had failed, we would have lost a major share of our scientific results.

On landing day the weather at Kennedy Space Center was not good. Normally, we would have spent another day on orbit hoping that the weather would clear and we could land at Kennedy. But the German mission manager was so worried about the freezer and its contents that we landed on the planned day, but at Edwards rather than Kennedy.

During ten days in space, we completed all of our experimental goals. I never slept more than five hours any night. When we landed, I was exhausted, but we continued to have tests done on our bodies for two weeks after the mission to see how we readapted to one-g. I got very tired of being a lab rat!

I barely had time after the testing was complete to visit my dad. He was not doing well and had been in and out of the hospital. Dad was seventy years old, and like many of his generation, he had been a heavy smoker. He had emphysema and congestive heart failure. It was very difficult to see this strong guy who had worked hard his whole life get to the point where he was fighting for every breath. I don't think Dad shared my interest in space, but he was always one of my biggest supporters.

After visiting him, I came back to Houston and packed for Germany. The entire STS-55 crew and our spouses spent two weeks traveling throughout Germany, giving presentations about our mission at universities and other forums across the country. We finished our tour in France at the Paris Air Show, and the rest of the crew was heading home the next day. Karen had never been to Europe before, so we planned to spend a few days in Switzerland; however, in the middle of the night I received a call from ASCAN classmate Dave Leestma telling me that Dad had passed away. Karen and I changed our airline tickets and headed back to Indiana.

I stayed in Indiana after the funeral to help Mom with all of the paperwork and decisions she faced. She was very strong and had done everything she could to take care of Dad, just like she had for both of her parents.

While at Mom's, I received a call from Hoot Gibson, who had recently taken over as the Chief of the Astronaut Office. Hoot told me that, in addition to being the Chief of the EVA Branch, he wanted me to serve as his Acting Deputy Chief of the Astronaut Office. Linda Godwin, his deputy, was being assigned to a flight and needed to start training. I served as the Acting Deputy Chief until after Linda finished her STS-59 activities in late summer 1994.

Most of my duties as Acting Deputy Chief were mundane daily office chores, but I did have the opportunity to attend a most interesting meeting in February 1994. I met with Academy Award winners Brian Grazer, Ron Howard, and Tom Hanks. They were doing research for their upcoming movie *Apollo 13* and were looking for advice and support for their project. It was fascinating to meet a group of professionals in a completely different field from mine and see that they approached their work with the same intensity and enthusiasm. The movie was partly filmed at the Kennedy Space Center and on a set in a zero-g KC-135 flown from Ellington Field near Johnson Space Center. *Apollo 13* was a huge success, grossing over $355 million worldwide, receiving nine Academy Award nominations and winning two Academy Awards.

Later that year in October, Karen and I had one of the most thrilling moments of our lives. Our daughter, Amy, was at Purdue University studying mechanical engineering, and Karen and I flew to Indiana to spend the weekend with her on the Purdue campus and see a Boilermaker football game. We checked into the Purdue Union Club Hotel relatively late, and we were walking down the main hall of the Union. It was unusually quiet; there was nobody else there. From around a corner, dressed in tuxedos, appeared Neil Armstrong and Gene Cernan. Simultaneously they saw us, raised their arms, waved, and called out, "Hey, Jerry, how are you doing?" For Karen and me to be back on campus at our alma mater and have two of my heroes—fellow Purdue alumni and the first and the last men to walk on the Moon—instantly recognize me and greet us in such a fashion was a special moment we will never forget!

Neil Armstrong was a class act. He was an unassuming, soft-spoken, thinking man, who almost seemed shy until one spent some time with him. I always respected the manner in which he carried himself and the strong principles that guided his life. I am proud of our common Boilermaker heritage. I am delighted to have had several opportunities to share time with Neil and his family. Most of these were various events on the Purdue campus. Neil had a wonderful sense of humor and just enjoyed being a regular guy. He believed in a strong America and in an ambitious American space program. I admired Neil's willingness and effectiveness in continuing to serve our country's space program in diverse and numerous ways. As Vice Chairman of the Rogers Commission that investigated the *Challenger* accident, he helped NASA return to flying the Space Shuttle as safely and as quickly as possible. Neil's testimony in numerous congressional hearings obviously carried significant influence. He was a strong advocate for challenging the American human spaceflight program with meaningful goals and giving NASA adequate resources to pursue those goals. I don't think America or the world could have chosen a better person to be the first human being to step onto the Moon. I was stunned when I heard of Neil's death on August 25, 2012. Neil read a draft of this book shortly before he died and offered technical comments and encouragement. With Neil's passing, the world lost a humble historymaker. I lost a hero and a friend. I will never forget what an inspiration and a true gentleman he was. We will never forget what Neil did for America and for mankind.

Gene Cernan is animated, intense, gregarious, and has a great sense of humor. Like Neil did, he continuously advocates for a vital US space program with aggressive goals to take us back to the Moon and then on to Mars. He frequently teamed with Neil to testify to congressional committees or to write white papers calling for a stronger US space program. When I read Gene's autobiography, I was surprised to learn that we had similar childhood experiences. Gene, a big-city, Chicago-land native, spent a great deal of time during his summers working on his grandfather's farm. I have really enjoyed talking to Gene about the early years of spacewalking development and training at NASA. The capabilities we now have to use spacewalks to build space stations and repair satellites in orbit are a direct result of the work that he and many others did in the earlier years of our space program.

The Soviet Union collapsed in 1991, and by the end of 1993, just months after my Spacelab mission, the United States, the Russians, the European Space Agency, Japan, and Canada announced plans to build an International Space Station. A space station was what we had been working toward. We were excited, determined, and very busy preparing for this ambitious international project. There was a lot to do, and it was back to work on EVAs.

Additionally, Russia and the US agreed to a series of joint flights. US astronauts would fly long-duration increments on the Russian Mir space station, and Russian cosmonauts would fly missions on the US Shuttle. Normally, there were three people stationed on Mir, and now one of them was going to be an American astronaut. During the joint program, seven US astronauts spent months at a time on Mir.

In June of 1995, flight STS-71, the third mission in the Shuttle-Mir program, accomplished the first Shuttle docking with Mir. In November, Shuttle mission STS-74, the fourth flight in the joint program, docked with Mir for the second time, and I was on board for my fifth flight into space. The purpose of our mission was to add a long, Russian-built docking module to Mir to make all future Shuttle dockings easier and safer.

Ken Cameron, my friend from STS-37, was given the command of STS-74. He requested that I be assigned as his flight engineer, mission specialist two. During STS-37, Ken told me I had a black belt in spaceflight. I guess he meant it! Jim Halsell served as the pilot. As an Air Force test pilot, Jim had flown flight tests on the SR-71 airplane. My fellow mission specialists were Bill McArthur, an Army helicopter test pilot, and Chris Hadfield, a Canadian Air Force F-18 fighter jock and test pilot. This would be the second spaceflight for Jim and

Bill, and the first for Chris. This all-military crew was a bunch of mul-titalented, fun guys. We thoroughly enjoyed training, traveling, and flying together.

Since we would be working with the Russians and visiting their space station, our trainers logically decided we should learn some of their language. Ken Cameron was already fairly fluent in Russian. Bill and Chris are pretty quick studies in languages and were paired together.

That left Jim and me. We were paired for training in the beginner Russian language class. Beginners we began, and beginners we would remain. It seemed that Jim missed about every other class, all for apparently valid reasons, so we would repeat lessons to help him catch up. I made almost every class, but I needed to repeat the lessons, too. We were making little headway. Thank goodness, we ran out of time and the language training was dropped before we hurt our brains too much!

We were carrying Russian-built hardware to Mir, so a trip to Russia was in order. The whole crew traveled to Moscow, and some of our NASA trainers and flight controllers accompanied us. We trained in Moscow at the Energia Design Bureau on the docking module. We also went to Star City and met the crew that would be on Mir when we docked.

During our visit to the Energia plant where the docking module was being built, we were given a tour of the Energia Museum that held exhibits dating back to the start of the Soviet space program. I had visited the museum once before in 1990 during my Association of Space Explorers tour. Our guide at the museum was a young lady who was wearing a light green, fluffy angora sweater. One of our trainers called her "Fifi," and he pointed out a museum photograph taken during my previous visit to the museum. Fifi became extremely excited when she recognized me in the photo. For years after that, everyone who was on that trip would ask me how Fifi was doing.

STS-74 launched on November 12, 1995. On the third day of the mission, Chris Hadfield, a Canadian Space Agency astronaut, used the Canadian-built robotic Canadarm to pick up the docking module from the payload bay. He carefully lowered the module and attached it to the docking system that had been extended from the top of the

Shuttle airlock. Once the module was installed, we opened the hatches between the Shuttle airlock and the docking module. We entered the module and installed a TV camera looking out the window in the hatch at the far end of the module. We used the view from this TV camera to guide us into a docking with the Mir station.

Mir was large, had multiple modules, and had solar arrays all over its exterior. We could see evidence of some damage from micrometeorite hits on those arrays. We had been forewarned to expect offensive odors in Mir due to long years of occupation and the growth of mildew fostered by moisture condensing behind panels. When we entered Mir, we were pleasantly surprised by the lack of offensive smells.

We did, however, feel like we were entering someone's crowded closet. There were bundles and bags of supplies and equipment stored everywhere. The Russians had the capability to carry more equipment and supplies to the station than they could take away, and the crews had to find places to stow the excess. After a while, well before we arrived, they ran out of good places to store the materials. They started putting bundles of stuff in the translation paths between the various modules of the station.

The crew on board Mir included Commander Yuri Gidzenko, whom I met on my first trip to the Soviet Union in 1990, when he was in training and evaluation as a cosmonaut candidate. The other crewmembers were Russian Sergei Avdeyev and German Thomas Reiter, whom I first met when I was training for the German Spacelab mission.

We enjoyed our time with the crew on Mir. We were given tours throughout the station and were even allowed to enter the attached Soyuz crew-return spacecraft. The Soyuz capsule was a cramped volume even without space suits on. We gave the Mir crew a tour of the Shuttle. They thoroughly enjoyed looking out of the Shuttle's windows and seeing their station from a totally different vantage point. It was really neat to visit another space vehicle and its inhabitants!

We shared dinners in each other's spacecraft. One of my biggest surprises was seeing two Russian Orthodox icons on the wall over the dining table in Mir, right next to a picture of Yuri Gagarin.

Our crew left some items on Mir. We left supplies for American astronaut Shannon Lucid, who would shortly serve as a long-duration crewmember. Shannon loves brightly colored "neon" socks. We brought several pairs and "salted" them among the supplies for her to find like Easter eggs. I discovered that Yuri Gidzenko liked American country music, so I left one of my country music audiotapes for his listening pleasure.

On this flight we carried a copy of the "United Nations Space Agreement on the Rescue of Astronauts, the Return of Astronauts, and the Return of Objects Launched into Outer Space" that had been adopted in 1968. We took it across to the Mir station and held a small ceremony.

After our missions were over, both the Shuttle and Mir crews were invited to the United Nations. We met in New York City and returned the flown "Agreement" during a very nice function at the UN. The UN Special Advisor for Public Policy Gillian Martin Sorensen was our gracious hostess. Her husband was Ted Sorensen, the political strategist and speechwriter for President Kennedy. Ted was associated with the speech in which President Kennedy said, "I believe that this nation should commit itself to achieving the goal, before this decade is out, of landing a man on the Moon and returning him safely to the Earth."

Gillian arranged an excursion around New York City for all of us. Tim and Nina Zagat, the cofounders of the "Zagat Restaurant Guides," served as our escorts. Our tour all over Manhattan included stops at the World Trade Center and the trading floor of the New York Stock Exchange. For lunch they took us to a delicatessen, and Woody Allen was sitting at a table eating with friends. Tim Zagat took us over and introduced us. Tim seemed to know everyone. I enjoyed meeting Woody Allen, but it didn't seem that Woody was overly impressed by us.

From New York City our combined crews went to Washington, DC. After most Space Shuttle flights, the crews were invited to Washington by NASA Headquarters. These trips normally included briefings to NASA employees at Headquarters, possibly a visit to the White

House, and visits to offices of members of the House of Representatives and the Senate.

On this trip, one of our stops was a visit with Senator John Glenn—the first American to orbit the Earth—in his Capitol Hill office. I've visited with John Glenn on several occasions over the years and enjoyed the opportunities to share space stories and to get to know one of my childhood, and adulthood, heroes.

John flew on Space Shuttle mission STS-95, which launched on October 29, 1998. I flew the very next Shuttle mission, STS-88, which launched on December 4 that year. During the preparations for those flights, we spent some time staying in the Kennedy Space Center Astronaut Crew Quarters. One evening I had an amazing experience. Just the two of us were sitting at a table in the dining room. I remembered playing hooky from school to watch John's historic Mercury flight on TV. And now there I was talking to John Glenn about our upcoming flights, about how much space travel had changed since his first flight, and where we each hoped our country's space program would go in the future.

On July 18, 2003, I participated in the 100th Anniversary of Flight activities in Dayton, Ohio, and shared a stage with John. I was delighted to help him celebrate his eighty-second birthday, and everyone there joined in singing "Happy Birthday." John and his wife, Annie, are both very warm and welcoming, and just plain nice people.

Even though my work at NASA meant I was frequently away from home, I found as much time as I could to be with Karen, Amy, and Scott. Amy and Scott grew up surrounded by the space program. They were ages seven and six when we moved to Houston, and Amy was twenty-four and Scott was twenty-three when STS-74 landed.

Scott liked anything with wheels on it from the time he was a baby. When he was getting ready to turn sixteen, like all teenagers, he wanted his own car. Since he liked classic cars, we looked around and found a 1957 four-door Chevy Bel Air. The owner had stripped it apart to fix it up and then didn't have time to work on it. I bought the car. We towed the larger parts into our garage and brought the rest home in boxes, jars, and baskets.

At every free moment, we worked on that car. We sandblasted and primed the entire vehicle. I continued to get sand out of my hair and ears for years after. When we finished, the car looked and ran great. I served as the chauffeur when we used it to take Scott and his bride, Faith, from the church to the reception after their wedding. Scott still has the car.

———————————————●———————————————

SCOTT ROSS: *Dad figured if I put some labor into a car, I'd be less likely to go out and tear it up. And he was right. We were looking for an old car or truck, and we came up with this '57 Chevy. It was quite a project.*

People talk about basket cases—this was the basket case. It literally came in baskets and boxes. But I learned a lot by fixing it up, and that car means a lot to me. Dad and I enjoyed doing it together. I know every nut and bolt in it. I could rebuild another '57 Chevy tomorrow if I had to, but I hope I never have to again.

The car has won some trophies and awards. The paint is what really catches people's eyes. I actually had that color picked out from the time I was eleven years old. I had seen that color in a magazine and wrote down the name: Du-Pont Imron Sovereign Blue. I loved it. And when we finally got the car, I knew exactly what color it would be.

Restoring my '57 Chevy was actually fun work, and I ended up starting an antique car club when I was a student at Texas A&M. The club is still active.

The car is still going great, too, but I don't drive it much. I'm very careful with it because it means so much to me.

I got engaged in that car, and Dad drove Faith and me away from our wedding in the car.

When we were looking at buying our house, Faith said, "Why don't you sell that car and get some money? We'll need it for the house." I told her the car was not going anywhere: "My dad and I did it together."

Dad never drove a fancy car himself. You might think astronauts all drive Corvettes, but not my dad. We weren't a two-car family until after we moved to Houston, and then he got a beater of a Toyota Corolla for himself. It just had a spray can paint job. Next he got a Toyota pickup truck. It ended up being mine while we worked on the Chevy. That Toyota had been in a wreck, and you had to turn the steering wheel a little to the left to keep it going straight down the road.

I bought a new truck when I graduated from college. At first I thought something was wrong with it. I couldn't drive it straight. I was scared to death to drive it because I was so used to holding the steering wheel over to the left.

Dad always drove whatever older car we had to work, and Mom had the nicer car. To tell you the truth, my dad is the most frugal person I've ever known and also probably the most down-to-earth.

Just recently he bought two new cars, and I almost had a heart attack. He's never done anything like that in his life. He's never been interested in a sports car. People have a predetermined idea of what an astronaut's life is like, and it's 100 percent wrong at our house.

When we finished the Chevy, Dad was getting ready to go on a mission, and he asked me what special thing I had that I wanted him to fly in space for me. I didn't have to think a minute.

I handed him the key for the '57 Chevy.

———————————————————●———————————————————

After Woody Spring and I demonstrated assembly techniques in orbit on STS-61B, it was very apparent to everyone that the Johnson Space Center water tank was not big enough. The Weightless Environment Training Facility tank we used for EVA training was thirty-three feet wide, seventy-five feet long, and twenty-five feet deep. It was not much bigger than the training mock-up of the Shuttle's payload bay, and at that time we were planning to build a much larger US Space Station named "Freedom." How were we going to train to build a huge space station in such a small tank?

Our STS-61B crew took the problem to Neil Hutchinson, then manager of the fledgling Space Station program. He understood. He agreed. "Why don't you guys go make it happen," he said.

Johnson Space Center formed a team of EVA experts and put together a briefing package to present at NASA Headquarters in Washington, DC. I became the pitchman and ended up going to Washington once or twice a year for about nine years, presenting our case to various review boards. Probably the only astronaut in the history of NASA who couldn't swim was leading the charge for a new pool.

The money for construction of new facilities comes from a different pocket than normal NASA operating funds. Funding for a new facility is a separate line item in the budget that goes before Congress for passage. The amount of money approved in any year is relatively small, and there's a lot of competition. We started making presentations in 1986 or 1987. A couple of times we were almost funded, and then something would happen and we were out.

In 1995, plans for the International Space Station were progressing rapidly. Mr. Abbey, then Director of the Johnson Space Center, understood that we needed a larger pool facility, and he didn't think we could wait any longer. With Mr. Abbey's assistance, Construction of Facilities funds were finally approved, and the new Neutral Buoyancy Laboratory was built in 1995-1996.

The new tank is 102 feet wide, 202 feet long, and 40 feet deep. The concrete floor is 6 feet thick. The walls range from 2.5 feet to 5 feet thick. It holds 6.2 million gallons of water and was slowly filled over a month to avoid cracking.

I was put in charge of leading the Operational Readiness Inspection and certification of the new facility. We looked it over from top to bottom. I did an inspection in scuba and also became the first astronaut to do a suited run in the tank.

In late 1996, we finished the certification of the Neutral Buoyancy Laboratory (NBL), and crews started training in it in January 1997. The

NBL is part of the Sonny Carter Training Facility, which was named for astronaut Sonny Carter, who sadly died in a commercial airplane crash in 1991, on the day my third flight, STS-37, launched.

We could simultaneously place a good share of the entire Space Station mock-up and the entire Shuttle payload bay mock-up in the pool for astronaut training. The International Space Station could not have been assembled without this facility. It was critical for developing the EVA hardware and assembly techniques for ISS and for training the crews who performed the on-orbit tasks. NASA also used this tank to train the astronauts who did all but the first of the Hubble Telescope repair missions. The tank is well equipped and can be used for training astronauts for future programs, maybe for missions to Mars.

Just after completing the certification of this tank, I was sent to Japan in 1997 to join a team of NASA and Japan Aerospace Exploration Agency experts to review and certify their new Weightless Environment Training System, a version of our EVA training tanks. Koichi Wakata was also on the team. Koichi, an affable and talented Japanese astronaut, had flown his first Space Shuttle mission one year earlier. He and I performed the first suited runs in the Japanese facility after it was certified.

During the weekend, Koichi took me on a tour of Tokyo. We visited the Tokyo Zoological Park, a Buddhist temple, the pearl district, and a neighborhood flea market, and we walked through a beautiful city park. Everywhere we went, young Japanese girls recognized Koichi. All the girls shyly came up to him and politely bowed, then went away giggling. I called him "The Man." Koichi just laughed.

Koichi and I played softball together on the Astronaut Office's team, and when we discovered there was going to be a baseball game that Saturday night, we made plans to attend. We arrived after the game had started, and Koichi decided to buy tickets to sit in the outfield bleachers. He certainly knew where the action was! The fans in the bleachers could teach the Wrigley Field Bleacher Bums a thing or two. It was obvious that these fans had been enjoying some good Japanese beer and maybe some sake. They were beating drums, blowing whistles and horns, and waving flags on long poles. They followed the game closely and were the cheerleaders for the rest of the fans. What a great place to watch the game!

In the summer of 2011, shortly after the last flight of the Space Shuttle program, I went to the Neutral Buoyancy Laboratory. Two astronauts were preparing for a training exercise. I ran into Steve Voyles, who has been around NASA for as long as I can remember. He takes care of the EVA training tools at the facility, making sure tools are ready whenever they are needed. Steve always gave me a hard time, saying I used too much force and broke his tools. He called me "The Bull." I admit, I'm not the most gentle person sometimes, but I told him it was better to find out if a tool was going to break here on the ground than up in space.

Years ago when I was training in the Weightless Environment Training Facility, Steve dropped pennies, nickels, and dimes into the tank. He'd taunt, "If you're so good, why don't you pick those up off the bottom of the pool with those thick space gloves?"

During the training run, I retrieved them. And I kept the money. He kept trying to figure out how I did it, because he didn't think it was possible to pick up coins with those stiff gloves and thick fingertips. He kept pitching in more pennies, nickels, and dimes. And I kept picking them up and keeping them!

One day I told Steve he was pretty cheap just throwing in pennies, so he threw in a Kennedy half dollar. I went into the pool, brought it up, and gave the coin back to him. Half a dollar was too much to pocket.

Finally, Steve used an underwater viewer to learn how I was recovering the coins. It was simple. I didn't try to pick them up with the stiff, bulky fingers on the gloves. I went to the bottom and waved my gloved hand over the coin. The force of the swirling water popped the coin up off the floor of the tank, and I grabbed it.

When I ran into Steve at the Neutral Buoyancy Laboratory after the last Shuttle flight, he reached into his pocket and pulled out the Kennedy half dollar.

"I've carried this every day since you gave it back to me," he said. "It's good luck."

When I retired from NASA, Steve came to the party. He gave me the half dollar and told me it was my turn to carry it for a while. I made many great friends at NASA!

After STS-74, I did something I had never done before. I turned down a possible opportunity to fly.

Bob Cabana, who was then Chief of the Astronaut Office, asked me if I would be the backup to Dave Wolf, another Purdue astronaut. Dave was getting ready to fly to Mir and stay there for months. He ultimately spent 128 days on Mir in late 1997 and early 1998.

I would only be the backup, and the odds that I would actually fly the mission were very small. I told Bob that I had been preparing my entire astronaut career to be the guy who helped lead the EVA effort to build the Space Station, and the International Space Station program was at the point where we were ready to start doing that. Everybody knew it was going to take hundreds of hours of EVA to build the ISS, and we didn't have that many trained crewmembers. I felt very deeply the best place for me was helping to build the new ISS.

In addition to that, I felt I had given NASA a lot on the D-2 Spacelab mission. I had spent many long months away from home and my family, and I wasn't ready to do that again while preparing for a Mir mission. Fortunately, Bob didn't argue.

However, this really could have gone badly. NASA is no different from any other organization. When you turn down an opportunity, you wonder if you will be offered any more. I stuck my neck out, no doubt about it, especially after five flights. There were not many people being offered a sixth flight, and no one had ever flown a seventh. My decision could have ended my career.

But Bob didn't hold it against me. In fact, about six months later he assigned me to STS-88, the mission that would start the International Space Station assembly. I was going to do spacewalks again. And this time, instead of doing two EVAs like I had on previous missions, we would be doing three.

I was elated! But little did I know that my astronaut career would soon be going to the dogs.

EIGHT

The John Young glass ceiling

Some astronauts have neat nicknames—Gus Grissom, Deke Slayton, Gordo Cooper, Buzz Aldrin, Hoot Gibson. On my sixth mission, STS-88, I picked up a nickname, but I am glad it didn't stick. Not only did I get a strange nickname, but to my surprise, my experience working on Scott's car came in handy.

Bob Cabana was our commander. Bob's first flight had been STS-53 in 1992, with Commander Dave Walker, a sandy redhead who was known throughout his Navy career as "Red Flash." Dave proclaimed the STS-53 crew would be "The Dog Crew" and altered his name to "Red Dog." The other crewmembers all took dog names too, like "Dog Face" and "Dog Gone." Bob was "Mighty Dog." The Dog Crew fixed up an old station wagon to look like a Space Shuttle, complete with a damaged T-38 wingtip as the Shuttle tail, and they called it the "Dog-mobile." The Dog Crew had a lot of fun with their dogginess.

"Red Dog" Walker next commanded STS-69 in 1995, and this crew called themselves "Dog Crew II." One of the astronauts rounding out that pack was physicist Jim "Pluto" Newman.

Bob "Mighty Dog" Cabana and Jim "Pluto" Newman were flying together with me on STS-88, the first International Space Station assembly mission. They decided they would like to continue the tradition with our crew, so we became unofficially known as "Stealth Dog Crew III."

Frankly, I wasn't too excited about the idea, but I went along. The old Dogmobile was still around, but it was beyond repair, so Rick "CJ" Sturckow, our rookie pilot, bought a similar model.

CJ is a highly disciplined and hardworking Marine, so he was aptly dubbed "Devil Dog." He and I did most of the work on the car, with Rick in the lead and me in the supporting role. Commander "Mighty Dog" and mission specialist and fellow spacewalker "Pluto" also worked on the car.

I was nicknamed "Hooch" because I am "lovable and loyal." Hooch was the name of the dog in the 1989 comedy movie *Turner and Hooch* starring Tom Hanks. I had never seen the movie, and I didn't find out until much later that Hooch was a slobbery dog that died at the end. This was not exactly the nickname or image I was hoping for!

It's not surprising that "Laika" would be selected as a nickname for a female astronaut. Laika was the name given to a stray female dog found on the streets of Moscow. She became the first animal to orbit the Earth in November of 1957, just a month after *Sputnik 1*. She died in orbit from overheating, but the flight proved that a dog, and therefore a human, could survive a launch and the zero-g conditions of space.

The final US member of our crew was Mission Specialist Nancy Currie, and Nancy was christened "Laika." Laika, being a smart woman, wanted no part of the dog car.

"Mighty Dog," "Devil Dog," "Pluto," and "Hooch" drove Dogmobile II to various events so we could arrive in style. We always offered Nancy a ride, but she always had an excuse and showed up later in her own vehicle. Finally, after yet another declined invitation of a ride, the rest of us boarded Dogmobile II and parked outside the exit of the building where she was working. We had a car door open and the motor running. This time when "Laika" came out, we had her. She went along with the fun and only barked at us a little.

When I was assigned to STS-88, I was only the third astronaut to be named to a sixth flight. The first was John Young, and no one had flown seven missions. I was one of NASA's most experienced astronauts. The problem with experience is that it comes with age.

To some, the word "experienced" is just a nice way of saying "old." I looked around the Astronaut Office and realized it was full of younger people, and they started kidding me about my "experience." "Hey, what was it like being an astronaut back in the Mercury days?" they'd ask. Or they'd want to know if I actually had to wrestle with the chimpanzee to see who would get the first flight. They asked which of my Gemini flights I liked the most. CJ was especially effective at teasing me and took great delight in it. While our son, Scott, was at Texas A&M, he found a 1939 Plymouth coupe that he convinced us to buy. It was a real classic and fun to drive, but CJ told everyone I was the original owner.

Cosmonaut Sergei Krikalev from the Russian Federal Space Agency joined the crew late in our training. Naturally, Sergei became "Spotnik." Spotnik was a great addition to the crew. This was Sergei's fourth flight, including fifteen months spent on Mir. In addition to his experience, having Sergei with us made working with the Russians a lot easier.

The differences between American and Russian cultures kept things interesting. Overall, Russian customs caused the Russian engineers to seem quiet, self-constrained, and usually not willing to share information. Their first answer to our questions was almost always "Nyet!" Sometimes it would be "Maybe." But they rarely said "Da."

Russians also deal with time differently than we do. Generally, they don't have an internal clock like Americans that says, "I've got five more minutes to get this done." Our training and preparations were schedule-driven, but the Russians didn't seem to proceed any differently if something needed to be done in five minutes or two weeks. Sergei was highly admired in the Russian space community, and we saw a different kind of cooperation when he joined the crew. He understood us, and he understood his countrymen.

The idea of a space station has been around since Jules Verne, who pioneered science fiction writing in the 1800s, and I remember very well when the United States started talking about building its own space station.

On July 4, 1982, I flew in a chase plane assisting the landing of STS-4 at Edwards Air Force Base. A large crowd had come to Edwards to celebrate Independence Day in a very patriotic way. Roy Rogers was among them, standing along the taxiway, and I was excited to meet my childhood cowboy hero. We exchanged autographs, and he signed mine "Many Happy Trails." President Ronald Reagan was there, too. In a speech after the landing President Reagan stated that he thought the United States should build a space station, and I was there to hear his words. The space station would be a place where astronauts would live and where vital research would be conducted. It would be a place where we could test and validate important capabilities that would eventually permit us to permanently live on the Moon and fly to Mars.

STS-88 was the start of the on-orbit assembly of the International Space Station (ISS), one of the largest and most ambitious international engineering and scientific endeavors in all of human history. The assembly would be a monumental construction task. Now completed, the ISS covers an area the length of a football field, including both end zones, and one and a half times a football field's width. It weighs 920,000 pounds. It has more living space than a five-bedroom house, and includes two bathrooms, two dining areas, two gyms, and a 360-degree bay window for spectacular views.

Our STS-88 crew made several trips to Russia to prepare for our flight. As we got ready, the Russians prepared to launch the *Zarya* module which would be the first piece of the Station sent into space. "*Zarya*" means "dawn," and *Zarya* represented the dawn of the International Space Station. *Zarya* was also called the *Functional Cargo Block*, which is much more descriptive of its purpose. It provides power, communications, guidance, propulsion, and storage space.

The plan was for the Russians to place *Zarya* into orbit using a Proton rocket. NASA would then launch the Shuttle *Endeavour* with the US-built *Unity* module, also called *Node 1*, in the payload bay, rendezvous with *Zarya*, and attach *Zarya* to *Unity*. Coupled in space, *Zarya* and *Unity* would become the cornerstone of the ISS. On one of our trips before STS-88, I again visited the Russian launch site in Kazakhstan. Our crew was trained on the *Zarya* spacecraft, and Jim Newman and I inspected the areas on its exterior where we would be working during three spacewalks.

Along with *Unity*, *Endeavour* would carry two small satellites in the payload bay, one for the US Air Force and one for Argentina.

Bob, CJ, and I went to Argentina during training for the mission to learn about their satellite. President Bill Clinton was also visiting Argentina at the same time. When the US Department of State learned there were three astronauts in the country, they wanted to include us in a ceremony with President Clinton at a resort in the Andes. NASA Headquarters told us we would be joining the event, and Bob was asked to introduce President Clinton. We weren't given much more information.

At the appointed time we showed up in our blue flight suits, and security just waved us right through. We met President Clinton and Argentinian President Carlos Menem, and Bob introduced President Clinton. Bob's introduction must have been well received, because in 1998, he was selected to sit next to First Lady Hillary Clinton in the US House of Representatives chamber gallery for President Clinton's State of the Union address. During the message, President Clinton spoke briefly about the International Space Station and also mentioned the upcoming Shuttle flight with Senator John Glenn on the crew.

The unmanned *Zarya* launched on November 20, 1998. Our whole STS-88 crew gathered at Bob and Nancy Cabana's home in the Houston area to watch the launch on NASA TV. We cheered as we saw the large Russian Proton rocket push *Zarya* toward outer space. We cheered again when it was announced that *Zarya* was safely in orbit.

Now it was our turn. We launched on the Shuttle *Endeavour* on December 4, 1998. It was my second night launch. Night launches were spectacular to experience from inside the orbiter or from many miles away on the ground.

On Flight Day 3, Nancy Currie used the Canadian-built robotic arm to lift *Unity* out of the payload bay and onto the Shuttle orbiter docking system. *Endeavour* had to dodge some space debris, but everything went according to plan.

The next day we rendezvoused with *Zarya*. Bob placed *Endeavour* within 33 feet of *Zarya* as we orbited the Earth, about 240 miles up, flying at 17,500 mph—leaving little margin for error. Next it was up to Nancy to operate the robotic arm and grab *Zarya*. Once she had grappled it, she lowered *Zarya* to within just inches of the mating adaptor on *Unity*, and Bob fired Shuttle jets to bring the two modules together. Now it was time for the spacewalkers to go to work.

On Flight Day 5, I floated out of the airlock on my first spacewalk since April 1991. Jim Newman and I electrically hooked the two modules together. The EVA lasted seven hours and twenty-one minutes, and I didn't have much time to take in the spectacular views. I only saw the Earth in passing as we worked.

Our second EVA was on Flight Day 7, and this was another seven-hour walk. Jim and I continued the work to connect the two modules, mounted two US communications antennas onto *Unity*, and dealt with a problem on *Zarya*. Two antennas on *Zarya* had not deployed automatically as intended; they were stuck in their stowed configuration. We manually deployed one of the Russian antennas by hitting it with a metal rod. We would free the other one on the third spacewalk using the same method. At one point, as Jim and I were mounting the antennas onto *Unity*, I was hanging on by the fingertips of one hand while manipulating a two-hundred-pound antenna with the other. I realized that this task was at the limits of what I could physically do, and I was relieved when the antenna was securely latched into place.

As we continued our EVA tasks, we passed over South America at night, and all of our communications channels were quiet. Suddenly, over our headsets we heard someone talking very rapidly in Spanish. It was a man speaking, he was excited, and I could tell he was doing a radio broadcast of a soccer game. From over two hundred miles above the Earth we heard, "GOOOAAALLL!" Then we flew past his ground-based radio station and returned to normal communication with Mission Control. As far as I know, this was the only time a Shuttle crew heard a commercial radio broadcast on orbit. I wondered where the game was, who was playing, and who won!

Seeing the first two elements of the ISS joined was impressive. *Zarya* is a little more than 41 feet long, 13 feet wide, and weighs 43,000 pounds. *Unity* is 15 feet wide, 34 feet long with the Pressurized Mating Adapters at each end, and weighs 26,500 pounds. One Pressurized Mating Adapter was mated to *Zarya* and the other was used for Space Shuttle dockings.

Flight Day 8, on December 10, 1998, was an historic occasion. Bob and Sergei led the way as the entire crew left *Endeavour* and floated into the ISS—the first time any human beings were inside it in space. As soon as Sergei entered the Russian module, he knew something was wrong. "Something doesn't sound right. It's too quiet," he said. And he was right. Sergei found a piece of plastic that had been left in one of the air ducts significantly blocking the airflow. We had a press conference from space after we entered the ISS and were excited by the significance of what we had just done. At the end of the press conference I grabbed CJ's head, pulled it over to me, and placed a kiss on his forehead.

"Aw, all my Marine buddies are going to see that," he grumbled.

I told him I was just celebrating the spirit of international cooperation in space. He should have been happy I didn't slobber like Hooch.

Navy Captain Mike Smith had been in my astronaut class. He was the only member of our class who died in the *Challenger* accident. While preparing for STS-88, I spoke with his widow, Jane Smith-Wolcott, and asked if it would be okay if I put one of our class patches on the new Station in memory of Mike. Jane thanked me for thinking of Mike and her family and said that they would appreciate the memorial. When we gained access to the ISS, I said a short prayer and velcroed a 1980 class patch to a wall behind some panels in the *Unity* node in memory of Mike. As far as I know it is still there.

The third and final spacewalk was on Flight Day 10. Jim and I continued our work on the modules. We climbed all the way to the top end of *Zarya* to install a handrail for future spacewalkers to use. While we were there, I noticed a problem with two Russian experiments on the outside of *Zarya*. They were about to float free from their mountings, so I requested permission to fix them. It took a very long time for the Russians to come back with an answer, but finally they said, "Da." Good thing. I had already fixed them.

I waited until this EVA to publicly comment on the Phase VI space gloves on which our daughter, Amy, had worked. With two engineering degrees from Purdue University, Amy came to NASA full time in 1996 and went to work certifying the new gloves.

I was testing the Phase VI gloves on STS-88. I was supposed to do the first EVA with the new gloves and the second with the old ones. I could then choose which gloves I wanted to wear for the third spacewalk. I never wore the old ones. I used the new gloves on all three spacewalks. And at the end of the EVA, I reported how good the new gloves were and said, "Tell Amy they work great!"

AMY ROSS: *Our home life in Friendswood had always been very normal. Sunday night was my favorite night. That was family night. We'd have a light dinner, make popcorn, and watch* Murder, She Wrote *on TV.*

Growing up with a father who is so excited about space, it rubs off on you. He sometimes took Scott and me to the office with him on Saturdays and let us work on the computers. It was always fun.

I went to Purdue like my parents. I wanted to work at NASA, and the best path to that goal was to study engineering. My parents really did encourage me to look around at several colleges, but I had heard about Purdue my entire life, and my grandparents lived nearby. When I visited the campus, I loved Purdue. The deal was sealed when I was offered a scholarship.

I started in the fall of 1989. During my freshman year, I was accepted into the co-op program, which meant

I would spend one semester at Purdue and the next working at NASA. The only co-op job I wanted was with NASA because being a co-op is the way to be hired by NASA after graduation.

I had been with my dad when he did public appearances, but it was a bit of a shock when he visited me at Purdue. At home he was just Dad. People ask me what it was like to grow up with a father who was an astronaut. It was just normal for us. It was the same as growing up with a father who was an electrician. Some fathers wire houses. My father flew in space.

But when Dad visited at Purdue, he was an astronaut, and everyone made a big deal about it. Most times when he came to visit me, he also came for a big dinner, a talk, or some kind of activity. Sometimes I wanted my parents to just come and visit me, but that rarely happened, and I understood.

I received my master's degree in mechanical engineering—like my dad—in 1996. I decided I didn't want to pursue a PhD; I wanted to work at NASA, and I got the job! I was doing advanced work in space suits when a colleague Bill Spenny called and said, "We've got a job we need done. Would you come and work on it?"

My mentor Joe Kosmo had developed new gloves for the space suits, and I spent the next two and a half years getting them through the certification process. Knowing my dad was going to be the first person to use them in space was very exciting and a little nerve-wracking.

I stayed in Florida during his flight when he was testing the gloves, sharing a condo with Grandma, Dad's mom. One night very late I was lying on the couch in the living room watching the last spacewalk on NASA TV. Grandma was asleep in the recliner. Unexpectedly, I heard my dad say how good the new Phase VI gloves were, and I heard him mention me by name. That made me feel really great!

Then the NASA Mission Commentator Rob Navias, who really does a great job, came on TV and announced to the world that what astronaut Jerry Ross was referring to were comfort gloves, worn under the space gloves, that his daughter, Amy, had knitted for him.

"Knitted!" I let out a yell. Grandma woke up and almost fell out of her chair, alarmed that something had gone wrong on the spacewalk.

It caught me totally off guard. I had a master's degree in engineering, I had spent two and a half years working on those space gloves, and Rob said they were something I knitted! I couldn't believe it. I almost crawled through the TV to grab him and say, "Those were space gloves, and I didn't knit them. I'm an engineer!"

After the mission I went back to my job at the Johnson Space Center. All of my understanding coworkers had prepared a greeting for me. There, stacked all over my desk, were balls of yarn and knitting needles.

I still have those needles. And people still kid me about knitting my dad's gloves.

On the third EVA of the flight we also tested a new piece of equipment called the Simplified Aid for EVA Rescue (SAFER). The SAFER was there to address a concern that is as old as the idea of walking in space. A spacewalker could accidentally become separated from his or her vehicle and float away to certain death. If an untethered spacewalker lost his grip and became separated from a structure, no matter how close, he could not get back on his own.

That is why foot restraints, handholds, and tethers are designed and so carefully placed to prevent it from happening. And that is why the fundamentals of keeping a safety tether always attached to the Space Shuttle or the ISS is a core part of our training. Every time EVA astronauts train in space suits in tanks, they practice tethering protocols and then practice some more.

The tethers NASA uses are steel cables with a hook on one end. The cable is wrapped on a reel that is attached to our space suit. We hook onto the orbiter, the ISS, or whatever structure we are working on, and as we move around, the tether unreels, allowing us pretty free motion as well as security.

With the exception of the Moon walkers, all US spacewalkers were tethered to their vehicle until 1984 when Bruce McCandless and Bob Stewart performed EVAs using Manned Maneuvering Units (MMUs). These rocket backpacks allowed Bruce and Bob to fly free from the Shuttle. The bulky MMUs were used on two more flights that year to rescue disabled satellites and were never used again.

When the Shuttle orbited by itself, the orbiter had the capability to fly over and retrieve an EVA astronaut if one should come loose and start drifting away. The commander and pilot trained for that task.

But when the Shuttle began docking, first at the Russian Mir space station and later at the ISS, it became obvious that in the event a crewmember accidently became separated from the structures, the Shuttle would not be able to undock in time to make a rescue. Likewise, if a spacewalker was working on the ISS when the Shuttle wasn't around, there would be no way to save him. I had worked on the development of the MMU and thought something like it, but a lot smaller and less cumbersome, was needed to add an additional level of safety for spacewalkers. Such a unit would give spacewalkers a self-rescue capability.

Brewster Shaw was the Space Shuttle program manager at the time, and I approached him with the idea. He thoroughly understood EVAs and the importance of what I was proposing; however, as program manager he was always balancing competing requirements against the program's budget.

Brewster got back to me and said the idea was good, but he really didn't think the program had the money to design, build, and test the proposed units. "Okay," I said. "But, Brewster, what are you going to tell Karen or some other spouse when their loved one is drifting off with his oxygen running out and his battery failing?"

We received the go-ahead to design, build, and test the units.

SAFER does make EVAs less dangerous. It is much smaller than the MMU and is mounted to the sides of the primary life support system on the back of our space suit. If a spacewalker were to become separated, he or she can deploy a handle and activate the system. The astronaut would use a joystick on the SAFER to fly back to safety. The unit has just enough thrust to get the spacewalker back to structure.

Just before finishing the third spacewalk, Jim and I went back into *Endeavour's* payload bay. Jim held my safety tether but gave me some slack and let me float free a short distance away. I activated the SAFER unit and tested it briefly. Some things didn't work exactly as we expected. After the flight these problems were fixed, and now spacewalkers have a means to rescue themselves should they inadvertently float free from the ISS. Fortunately, the SAFER hasn't been needed for a self-rescue yet.

Back in the Shuttle, after we undocked, I was back at the windows. During one night pass over Canada, several of us spied something out the aft windows of the orbiter. It wasn't very bright. It was whitish in color and seemed to change moment by moment. We turned down the interior lights to reduce the reflection on the glass and pressed our noses to the windows.

We realized we were flying right through an aurora borealis! Excitedly, we called the rest of the crew to the flight deck so they could see it too. The white curtains of the aurora were dancing all around us. We all agreed that we could not capture the soft glow with our cameras, but our eyes allowed us to enjoy this wonderful, ethereal display from its midst. We must have traveled through the aurora for about six minutes, and during the whole time all that could be heard from the entire crew were "Ooh," "Aah," and "Look at that!" What a great memory!

After laying the cornerstone for the ISS, we touched down at Kennedy Space Center and had another very special moment. Standing at the bottom of the stairs ready to greet us with NASA Administrator Dan Goldin were Apollo astronauts Gene Cernan, Jim Lovell, and Bill Anders.

My sixth flight had been historic. We had joined *Zarya* and *Unity* and begun assembly of the International Space Station.

But what was going to happen to my journey in space now?

I had flown six flights. Three other astronauts had flown six flights, but no one had flown more. There were a lot of new people in the Astronaut Office waiting for a chance to fly, and I had grown relatively old, in astronaut terms. I was fifty.

Even worse, the consensus in the Astronaut Office was that NASA would never let anyone fly more times than John Young. He had a total of six missions in three programs: two in Project Gemini, two in the Apollo program, and two in the Space Shuttle program.

I never set out to break John Young's record. As far as I am concerned, I can never surpass John, no matter what I do. John is my hero. Over the course of forty-two years of active NASA service, John did so many incredible things for our country and our space program. I don't think anyone can ever surpass what he's accomplished.

Besides, if you talk with his wife, Susy, she'll tell you John actually launched seven times: six times from Earth—and one time from the Moon! Most people forget John launched from the surface of the Moon to return home.

I'd never argue with Susy.

I didn't want another flight just to break a record; I wasn't ready to retire or go on to something else with NASA. I wanted to fly again. In my heart, I was an astronaut and a spacewalker, and there was an International Space Station to assemble. I wanted to continue doing what I loved.

By this time, former STS-55 crewmate Charlie Precourt was Chief of the Astronaut Office. After I came home from STS-88, I went to Charlie's office to find out if NASA had a John Young glass ceiling.

Charlie said he didn't have any restrictions, so I requested that he check with Mr. Abbey. I wanted to find out if there was something higher up limiting astronauts to six flights. Charlie came back and told me there was no ceiling, so I told him I wanted to stick around in the Astronaut Office and try for more flights.

He pointed out there were only so many flights, and there were a lot of qualified people. When I saw each bunch of newly hired astronauts walk in the door, I knew my days were numbered. But I wanted to try.

I was assigned as the Chief of the Astronaut Office's Kennedy Space Center Operations Support Branch. We did a lot of testing of ISS hardware. I had a team of astronauts routinely supporting those tests around the clock, seven days a week. I also led the Astronaut Support Personnel, the team that supported each of the Shuttle launches and landings. I helped set up the Shuttle cockpit for crews before launch and coordinated all other astronaut support activities at Kennedy.

In addition to my work at Kennedy, I continued doing developmental work for the ISS spacewalks and served as an instructor astronaut, helping to mentor other crews preparing to fly. I kept busy.

One afternoon in early 2001, I was at home. I had just returned from the Cape after a launch, and I was taking some time off to work around the house. The phone rang. It was the Astronaut Office, and I was asked to come to the Johnson Space Center. Charlie Precourt wanted to see me.

I thought there were probably only two reasons that Charlie would call me into the office. Charlie was going to tell me to put on my flight suit. Or he was going to tell me it was over.

I drove straight to the Johnson Space Center just like I'd done for over twenty years. A person thinks about a lot of things on a familiar drive like that when he knows life is at a turning point, but I wasn't worried. I trusted in whatever God had in store for me, and I had already been given so many wonderful opportunities. I couldn't ask for anything more.

Charlie greeted me when I arrived at his office, and he didn't waste any words. He said I had been assigned to another mission, STS-110. My heart leapt. It was an ISS assembly flight, so I was going to get a chance to go back into space and do two more EVAs. I thanked Charlie for the great news! When I left Charlie's office, the rest of the crew was standing outside the door ready to celebrate our flight assignment.

The commander was Mike "Bloomer" Bloomfield. This would be his third flight, the first as a commander. Mike had been captain of the Falcon football team at the US Air Force Academy under coach Bill Parcells. The pilot was Steve Frick, a US Naval Academy graduate. Steve had been stationed aboard the carrier USS *Saratoga* and flew the F/A-18 on combat missions during the Gulf War. He was a space rookie.

This was a fully booked flight. In addition to me, there were four other mission specialists assigned: Stanford graduate three times over Steve Smith, who was on his fourth flight; Stanford electrical engineering PhD Ellen Ochoa, also on her fourth flight; Navy Captain, MD and PhD Lee Morin, on his first mission; and Rex Walheim, another rookie. Like me, Rex had graduated as the Outstanding Flight Test Engineer at the US Air Force Test Pilot School, but in Class 92A. I had graduated seventeen years earlier in Class 75B.

Rex quickly started telling people, "Jerry Ross and I have six previous spaceflights between us!"

By this time, I was a grandfather. Our son, Scott, had two daughters named Cassidy and Katie, and while Lee Morin would be a rookie in space, he was also a grandfather. I did two spacewalks with Lee on STS-110, and the rest of the crew called us the "Silver Team" because of the color of our hair. They were being nice. It sounded a lot better than the "Gray Team." Or the "Balding Team."

As I prepared for STS-110, I knew that it would most likely be my last flight. Not because I wanted to quit. I definitely did not! But NASA had hired many new astronauts over the last three years, and they wanted to give all of them opportunities to fly. The handwriting on the wall had become unmistakable. I would be fifty-four years old when we launched. Since these would likely be my last spacewalks, I was going to enjoy them!

STS-110 was going to be a challenging mission, as all of the ISS assembly flights were. We were the initial flight in the last phase of construction of the ISS, and we were going to install the center section, called S-zero, of the long truss that holds all of the solar arrays for the US part of the Station.

I really looked forward to getting another look at the ISS. I had been working on it since the start of the Station program. During my spacewalks on STS-88, we had put the first two parts together, and no one was living there. A lot had happened since. Now the ISS was continuously inhabited, and looking at pictures it was hard to even find the first two pieces we had connected to start the assembly. We were also going to depart on our spacewalks from an airlock on the ISS, as opposed to the one on the Shuttle.

S-zero is forty-four feet long and weighs nearly twenty-seven thousand pounds. It contains the equipment to convert and distribute electrical power generated by eight large solar arrays that are mounted on the completed truss. S-zero also routes the cooling fluids used to maintain the proper temperatures for all of the equipment inside and outside the ISS.

On the morning of Thursday, April 4, our liftoff was delayed four days when a substantial hydrogen gas leak was discovered on the mobile launch platform during fueling of the large external tank. We busied ourselves marking time at the Beach House, running, riding bikes, watching movies, and studying. My family left me behind and went to Disney World! On April 8, we posed for pictures, ate lunch, and headed for the suit room.

Astronauts have very few traditions; however, most crews followed one in the suit room at the Kennedy Astronaut Crew Quarters just before walking out the door for the ride to the pad. I don't know how the tradition began, but it started sometime after STS-26 and was around for most of the Shuttle years. It might seem strange for the crew to be playing poker just before heading to the pad, but that is what we often did. A NASA manager dealt the cards to the crewmembers one at a time, five cards each, and the crew played a very special type of poker called Possum Fargo. The crewmember with the worst hand was eliminated. While watching the clock, additional hands were played until the commander was the player with the worst hand. Only then, having symbolically used up all of the commander's bad luck for the day, was the crew, according to the tradition, cleared to head for the pad.

When Commander Mike Bloomfield turned over that losing hand, everyone in the suit room raised a cheer, which signaled to those waiting in the hall outside that he had lost and the crew was on its way. It was time to "go fly!" All smiles, we lined up at the suit room door with NASA managers in tow and stepped out into the lights of the cameras for the trip to the pad that was always seen on NASA television and in photographs. And it was a good thing we played astronaut poker that day. During the last seconds of the countdown, there was a computer glitch in the Launch Control Center. After some discussion and quick work by the launch team, we were finally cleared for launch with only seconds remaining in our launch window.

As the solid rocket boosters ignited, the Shuttle jumped off the pad. Steve Smith and I were seated next to each other in the mid-deck, and we gave each other high-fives. At that moment, I became the first human in history to launch into space seven times. But this launch didn't feel any more special to me than the others. Every Shuttle launch was spectacular.

Records are good, but they shouldn't last long. If we don't keep breaking records, we're not progressing. I hope my record is broken, and I hope it's done by an American. However, given the current state of human spaceflight in our country, that is not likely to happen anytime soon.

It took us three days to link up with the ISS, and the docking went perfectly. We were the first guests the crew in the Station had seen in the four months they had been there, so they were very happy to greet us. During our visit, we hosted them for a meal with some real Texas barbecue, and to top it off we made s'mores with tortillas (to minimize crumbs), Hershey bars, and marshmallows! The only thing we needed to make it really feel like Texas was a campfire, but NASA frowns on that.

The day after docking, Ellen Ochoa and fellow US astronauts Dan Bursch and Carl Walz, who were living on the ISS with Russian Yuri Onufrienko, used the Station's robotic arm to lift S-zero from *Atlantis*'s payload bay and to position it on top of the American scientific research laboratory, named *Destiny*.

We had two spacewalking teams. Steve and Rex did the first and third walks, while Lee and I performed the second and fourth. The Silver Team conducted its first spacewalk on Saturday, April 13. It felt wonderful to be out in space again. Lee and I went to work on the truss. We connected two of the four legs that attach the entire truss to the ISS. While we were working, I reminisced about Woody Spring and me testing construction techniques on spacewalks during STS-61B. I told Lee, "On 61B, Woody and I said, 'Let's go build a space station,' and now you and I are doing it!"

We removed and stowed two large structures that had held S-zero in place in the payload bay of *Atlantis* during launch. Then we mated several electrical connectors. The spacewalk lasted more than seven hours, but we were very busy and the time passed quickly—too quickly.

At that point the ISS still wasn't all that big, but there was plenty of room inside for all ten of us. *Destiny* was about the size of a single-wide mobile home. The Russian Service Module, *Zvezda*, was a little smaller. In comparison, the Shuttle felt cramped.

On my first Shuttle flight we had no way to privately communicate with our families. On STS-110, we could call our families from ISS. There is a telephone up there, and you can dial home just like you do on Earth. I called several people. I called my mom, and I tried to call Karen, but when I couldn't reach her I left a message from space on our answering machine. Our communication capabilities had come a long way.

On April 16, our second spacewalk lasted six hours and thirty-seven minutes. Lee and I installed the handrail path that all spacewalkers now use to go from the airlock hatch to the truss. We installed large spotlights that illuminate the exterior of ISS and facilitate spacewalks and robotic operations during nighttime. We installed some large electrical circuit breakers, set up EVA tools for future spacewalks, and finished preparing the mobile transporter for future operations. It's a day I'll remember for the rest of my life.

It was my last spacewalk.

Approaching the end of the EVA, my feet were anchored in a foot restraint on the end of ISS's robotic arm. I had completed my EVA tasks, and the arm was going to take me back to the airlock hatch where we would conclude the spacewalk. At this point, I was just along for the ride.

The sun would be setting soon. I was looking west past the Russian part of the ISS. Below my feet was the Mediterranean Sea, and I was traveling at five miles per second to the east.

As we passed the fertile, green Nile River Delta, the sun set in a spectacular display of red, orange, yellow, and purple that painted the horizon above the curving silhouette of the Earth. I turned off my helmet-mounted lights and raised my sun visor. I surveyed the stars and watched thunderstorms over Israel and Lebanon. My crewmates inside the Station started to move the robotic arm. Now the darkened Earth was over my head and the stars were below my feet. The ISS and portions of the Shuttle illuminated by lights passed in and out of my view as the arm continued to move.

I went past the Shuttle's vertical tail and felt like I could have touched it without much effort. I looked down into the empty payload bay where the S-zero had been stowed at launch. Just as the sun started to rise, the arm brought me past the starboard side of *Destiny* and underneath the S-zero truss that we had attached during the mission. I visually soaked it all in and made it a permanent memory to cherish forever.

The arm stopped at the ISS's airlock. I configured my safety tethers, removed the foot restraint from the robotic arm, and stowed it on the exterior of the airlock.

I took one final look at the ISS, at the Shuttle, and one last unobstructed glance at God's beautiful Earth. I reluctantly entered the airlock and concluded my last spacewalk. It was a melancholy moment.

I carried my Bible with me on my last flight. On earlier Shuttle flights we were not allowed to carry a personal item as large as a Bible. But this time I took my Bible with me and read it at night. One of my favorite verses is Philippians 4:13. I learned after my mother's death that it was also her favorite: "For I can do everything God asks me to with the help of Christ, who gives me the strength and power." I read that verse in space and thought about the journey I had taken. I read the creation story in Genesis while preparing to go to sleep in the ISS: "When God began creating the heavens and the earth, the earth was at first a shapeless, chaotic mass, with the Spirit of God brooding over the dark vapors. Then God said, 'Let there be light,' And light appeared."

I have no problem combining science and my faith; they are not mutually exclusive. I know and understand what science says about the creation of the universe and our world, and I believe it. Thanks to the US space program, we're going to learn even more. I also believe God had His hand upon it, and reading the creation story in Genesis while looking down on the beauty of the Earth from space was a very powerful moment for me.

Personally, I find it impossible to believe that everything I saw from space was created without God. Science explains many things. We understand more about the universe every day. But there are things that science will never be able to explain. To me, that is evidence of God's infinite wisdom and knowledge, and those things beyond our understanding are the things we can hope to have revealed in the life after death.

I really didn't have a lot of time to read during that last mission, but I wanted to make time for that experience of reading the Bible in space. And as I read, the words had special meaning because in my heart I knew this flight would most likely be the last time I would ever see the world from space.

After the flight, the STS-110 crew and our spouses were invited to the White House by President George W. Bush. We were greeted by President Bush in the Oval Office, and he was very warm and gracious, like his father. He was actually one of the nicest people I've ever met, acting relaxed and making us feel welcome and even at home in the Oval Office. As we walked in, on the wall to the right there was a painting that was very important to President Bush, and he told us why he wanted it in the Oval Office.

It was painted by W. H. D. Koerner in 1916 as an illustration for a short story in the *Saturday Evening Post*. In 1918, the title *A Charge to Keep* was attached to it by *Country Gentleman* magazine. The painting features a man on horseback riding at full gallop over the crest of a hill with a group of riders making their way up the hill close behind him.

When President Bush looked at that painting with its title, *A Charge to Keep*, he saw a Methodist circuit rider preacher with his companions. It reminded him of the hymn by Charles Wesley "A Charge to Keep I Have." The lyric of the second verse of the hymn says: "To serve the present age, my calling to fulfill; O may it all my powers engage to do my Master's will." President Bush felt those words described his own mission as he led our nation. The heartfelt way President Bush told that story made a deep impression on me. After he left office, President

George W. Bush wrote a memoir of his journey to the White House. It was titled *A Charge to Keep.*

After STS-110, Charlie Precourt told me it was "very unlikely" I would be assigned to another flight. In fact, he told me that NASA management actually had second thoughts about giving me, and later Franklin Chang-Diaz, our seventh flights. But they decided to go through with it because our crew assignments had already been announced. I told him I understood, but I would stick around the Astronaut Office, continue to work, and if an opportunity did present itself, I'd be ready to go. They hadn't totally shut the door, and there was always an outside chance.

Just over nine months after my seventh mission landed, the Shuttle *Columbia* came apart fifteen minutes from landing, killing everyone on board. I was at the Kennedy Space Center on the runway waiting to greet the crew, all friends, when I got the word.

If I had thought about it at that moment, I would have known flying in space was over for me, forever.

But that was the least of my concerns.

NINE

"Lock the doors"

On February 1, 2003, I was standing by the runway at Kennedy Space Center waiting for the Shuttle *Columbia* to land. It was just before nine o'clock in the morning, and even though the orbiter was still over Texas, it was only minutes away from its scheduled touchdown.

Since my last flight I had taken a new job as Chief of the Vehicle Integration Test Office. This job included managing the Astronaut Crew Quarters at Kennedy and taking care of the crew while they were at the Cape before launch and after their landing. *Columbia*'s mission, STS-107, was my first in the new job.

The day before launch, I had been in *Columbia*'s cockpit making sure the switches were all in the right position. On launch day, I rode part-way to the pad with the crew. I was among the last people to see them off, and on this morning, I was going to be one of the first to greet them upon their return from a successful and long—sixteen-day—mission.

Bob Cabana, my friend, former Chief of the Astronaut Office, and then the Director of Flight Crew Operations, was with me. We were standing outside the highly modified motor home that served as the landing convoy commander's command post. All activities on the runway after the Shuttle landed would be directed from inside this vehicle. The crews' families and their astronaut escorts were over at midfield beside the runway, waiting for *Columbia*'s return.

One of the guys in the command post vehicle waved for us to come inside. He said Mission Control in Houston had lost communication with *Columbia*. While not normal, that could happen. We waited quietly. Soon there was another report that Houston had lost data. That could also happen. If communication was lost, data also probably would be lost. The next piece of information came very quickly. Mission Control had lost tracking.

I left the command post and began to deal with our worst nightmare.

An hour earlier at 8 a.m. EST, Entry Flight Director LeRoy Cain in Houston had polled his team on a Go/No Go for re-entry. It was a Go. *Columbia* was coming down.

At 8:15 a.m., the orbiter was upside down and traveling tail first over the Indian Ocean. *Columbia*'s rocket engines fired to slow it down and start its entry into the Earth's atmosphere.

At 8:48 a.m., a sensor on the leading edge of the left wing recorded a high level of strain. This information was recorded on board but not telemetried to the ground.

At 8:53 a.m., *Columbia* passed over Sacramento, California. Sightings of debris being shed were reported by people out watching the Shuttle pass overhead.

At 8:54 a.m., flight controllers in Houston Mission Control began reporting readings from *Columbia* that were "off scale."

At 8:59 a.m., it was reported that pressure readings had been lost on both left main landing gear tires. Mission Control notified the crew that they were aware of the problem and were evaluating the situation. Commander Rick Husband responded: "Roger, uh, but . . ." Communication was lost.

At 9 a.m. came the loss of signal in Mission Control, and ground observers in Texas reported seeing multiple streaks of debris slashing through the morning sky.

At 9:12 a.m. in Mission Control, Leroy Cain gave the order to lock the doors and to make certain all information was preserved and kept in the room.

When I stepped outside the van, I said a quick prayer for the souls of my friends, and then I called the astronauts who were escorting the crew families. I told them we had lost tracking of the vehicle and to get the family members back to Astronaut Crew Quarters as soon as possible. The families were outside their bus lined up along the runway waiting for landing, and it was hard to gather them together. The escorts didn't tell them what had happened, but the adults knew from what the escorts were telling them to do and from the expressions on the escorts' faces.

I called the Astronaut Crew Quarters and talked to Lauren Lunde and Judy Hooper, who worked for me there in the office. I told them what had happened, to get security there, and to lock down Crew Quarters. I asked them to turn off all the TV sets and told them we would bring the families to the conference room. Finally, I grabbed Bob Cabana, the nurses, flight surgeons, and fellow astronauts who were out on the runway waiting for *Columbia*, and we all rushed to Astronaut Crew Quarters.

We beat the family bus back, and when it arrived, I helped escort the families to the conference room. I had already seen video on TV of the Shuttle breaking up over Texas and debris screaming across the sky, but I didn't say anything. The family members were in a state of shock. They asked me what I knew, and I told them NASA was trying to find out what had happened. I made calls and started to make arrangements to fly the families back to Houston. Finally, Bob Cabana came into my office, said he had received enough information, and that we should tell the family members.

We entered the conference room where the families were waiting. Everyone looked toward us, and Bob told them as gently as he could that the crew was gone. The vehicle had been destroyed during reentry, and there was very little likelihood that anyone had survived. We did what we could to comfort the families.

After the *Challenger* disaster, the Astronaut Office had written a contingency plan for dealing with an accident, but the plan was based on something happening at launch, not on reentry. We roughly followed the plan and otherwise used common sense.

During the next several hours, I sent people to pick up the families' baggage at the hotel and arranged for three airplanes to fly the families back to Houston along with their astronaut escorts, other astronauts, and medical personnel. The only pause in activity occurred when President George W. Bush called to offer the families his condolences.

Before I put the families onto the airplanes and said my good-byes, I was informed that the NASA Accident Investigation and Rapid Response teams were being activated and sent to Shreveport, Louisiana. I was a member of the Rapid Response Team. I wasn't going home.

It was hard to close the door of that last plane. I wanted to be on the airplane with the families, and I wanted to be on the flight back to Houston so I could be with mine. Karen knew the crew and their families, and I knew she and I shared the same pain. I wanted to be with her.

Driving back to Astronaut Crew Quarters, the place where I should have been celebrating with the exultant crew and the happy members of their families, I pulled off the road and let it go. I finally cried.

When I got back to Astronaut Crew Quarters, it was very strange to be there alone. I had been there with the crew before they launched, and I had been with their loved ones as the reality of what had happened sank in. Now Crew Quarters was empty; everyone was gone. I gathered the few clothes I had, put them in a bag, and went to catch an Air Force C-141 that was taking the Rapid Response Team to Barksdale Air Force Base in northwest Louisiana.

It was after ten o'clock at night when we arrived at Barksdale and after midnight when I finally reached my motel. I had been up since three o'clock the previous morning to prepare for the return of *Columbia*. It had been a long day, a day that I will never forget.

Before the launch, because of bad weather, the entire STS-107 crew had flown to the Kennedy Space Center in one of NASA's Shuttle Training Aircraft, a highly modified business jet, instead of flying there themselves in T-38s. I rode with them on the flight. They were a great crew and wonderful people.

Rick Husband: Rick was the commander. He was handsome, easygoing, a good singer and guitar picker, and a great family man. He was from Amarillo, Texas, and took over my duties as the "hometown" astronaut in Amarillo where I had been previously "adopted." He liked to say that his name was Rick Husband, not a Rich Husband, as his wife Evelyn would have preferred. Evelyn just rolled her eyes. Her book *High Calling* is an excellent account of their life together.

Willie McCool: Willie was a Naval aviator and a runner. He was the captain of the US Naval Academy's cross country team his senior year and graduated second in his class. Willie was always energy in motion and had all-American wholesome looks and manners. He always tried to leave anything he touched better than it was before.

Dave Brown: Dave worked for me supporting launches and landings at the Kennedy Space Center, and he had helped set up the cockpit and strap in crews for their launches. He was a Naval aviator and also a Navy flight surgeon. He was enthusiastic about whatever he did. I frequently flew back and forth to the Cape with Dave in NASA T-38s and always enjoyed our conversations about his wide range of interests. Dave was an accomplished photographer and videographer, and some of the images he captured prior to his only flight were used in a very touching production after his death. As his immediate supervisor, I had been impressed by his hard work and dedication and had encouraged the Astronaut Office management to assign Dave to the STS-107 mission.

Mike Anderson: Mike was a quiet, capable person who enjoyed his wonderful family. Mike had worked for me at the Kennedy Space Center for a year or so. He led the astronaut support for a very large and complex series of tests involving many of the major components of the US portion of the International Space Station. These tests were called the Multi-Element Integrated Tests. As our lead participant, Mike had to ensure Astronaut Office participation was available seven days a week, twenty-four hours a day to support the many crew-critical activities. I had also encouraged the Astronaut Office management to assign Mike to the STS-107 mission as the Payload Commander. I had flown to the Kennedy Space Center in a T-38 with Mike for the Terminal Countdown Demonstration Test, the practice countdown for *Columbia*.

Kalpana Chawla: KC, as she was known, was a very pleasant, soft-spoken, but energetic woman. KC not only had a PhD in aerospace engineering, but also Commercial Pilot licenses and Certificated Flight Instructor ratings. She always seemed to have a smile on her face and a gleam in her eyes. Since I had never worked directly with her, I didn't know her as well as I knew some of the others; however, I found her engaging and nice to be around. We would talk while working out at the astronaut gym about what we were each doing. She was full of spirit and was thrilled with her involvement in the space program.

Laurel Clark: I didn't work directly with Laurel either, so I didn't know her well. Laurel was a medical doctor and a Navy Captain. She was pleasant, personable, and full of life, displaying a warm smile. She was very devoted to her husband and fellow Navy flight surgeon, Jon, and their son, Ian.

Ilan Ramon: Ilan was a very accomplished Israeli Air Force pilot. He had an air of self-confidence but not one of arrogance. For whatever reason, he and I felt comfortable together from the very first time we met. Ilan was soft-spoken, intelligent, and interested in everything and everyone around him. He enjoyed life and was happy to work out of the limelight. As the first Israeli citizen to fly in space, that was not always the easiest thing to do.

As with *Challenger*, the cause of the breakup of *Columbia* was thoroughly investigated, although we all had a pretty good idea of what had happened from the start. And as it was with *Challenger*, this accident was something that could have been prevented.

The liquid oxygen and liquid hydrogen contained in the external tank at launch were at extremely cold temperatures, -297 degrees Fahrenheit and -423 degrees Fahrenheit, respectively. The tank needed a very good layer of insulation on its exterior to keep the oxygen and hydrogen in their liquefied states. The development of the materials to be used for this insulation and for the retention of the insulation on the tank during flight was problematic throughout the Shuttle program. Every Shuttle flight experienced some foam loss from the external tank during launch.

As a result, on every mission there was some damage to the extremely fragile thermal protection tiles on the underside of the orbiter from pieces of the foam hitting the tiles during launch. The tiles on the underbelly of the orbiter and the reinforced carbon-carbon panels on the leading edges of the wings and on the nose cap were critical to the Shuttle's survival during reentry into the Earth's atmosphere. The tiles had to withstand temperatures of up to 2,300 degrees Fahrenheit, and the reinforced carbon-carbon panels had to survive even higher temperatures—up to 2,750 degrees Fahrenheit, hotter than volcanic lava.

On STS-112 in October 2002, two launches before STS-107, a chunk of foam broke away from the tank and hit an electronics box on the solid rocket booster attach ring, causing a dent four inches wide and three inches deep. After that mission, the situation was analyzed, and NASA decided to press ahead.

On STS-107 at 82 seconds after launch while traveling at 1,870 miles per hour, a briefcase-sized piece of foam broke away from the tank, striking *Columbia's* left wing leading edge reinforced carbon-carbon panels. As demonstrated by ground experiments conducted after the accident by the *Columbia* Accident Investigation Board, this likely created a six- to ten-inch-diameter hole in that wing. During reentry at the end of the mission, extremely hot gases entered the wing through that hole, melting everything in its path, destroying *Columbia,* and killing the crew.

In both the *Challenger* and *Columbia* accidents, NASA management's decision-making processes were identified as contributing factors. In both incidents, NASA had observed on multiple flights prior to the accidents the very hardware problems that ultimately destroyed the two Shuttles and killed fourteen crewmembers. Since these specific hardware problems had occurred on multiple previous flights and those missions had safely returned to Earth, managers started to believe the hardware was more robust and could sustain more damage than the original design specifications allowed. Once again, it was "normalization of deviance," the term used by sociologist Dr. Diane Vaughan after the *Challenger* accident, to describe an illogical, inappropriate, and unsafe management approach.

Columbia debris was being discovered from Littlefield, Texas, just northwest of Lubbock, east all the way across the state and into Louisiana on the Fort Polk Army Base. Most of the debris was eventually found on a line from south of Palestine, Texas, to Hemphill, Texas. NASA spent three months searching 2.3 million acres collecting the debris.

About 40 percent of *Columbia*'s total dry weight was recovered and sent to the Kennedy Space Center. There the debris was set out on a large grid painted on an aircraft hangar floor in an attempt to reconstruct as much of the vehicle as possible. The recovery of the debris and the reconstruction effort were key elements in determining what had happened to *Columbia*.

The morning after the accident, I assembled a team of Shuttle engineers, and we traveled from Shreveport to Lufkin, Texas. It was Sunday, and Lufkin authorities wanted the team to sweep the school grounds to make sure there were no hazardous materials there when the children returned on Monday.

While the team went out to the schools, I stayed at the Lufkin Pitser Garrison Civic Center where the multi-agency Accident Command Post had been established, and I tried to start coordinating the resources that would be required to search for *Columbia*'s wreckage.

The Lufkin Civic Center was a beehive of activity. The thing that impressed me was that so many people could come together in one place so quickly. The United States of America has tremendous resources. We had communications equipment, trucks, airplanes, helicopters, and about one thousand National Guard members. There was an alphabet soup of federal, state, and local agencies already on site, and I was afraid it was going to be a nightmare to operate with all these different groups. I was wrong. Everyone came with their eye on the job, and the cooperation was wonderful. The outpouring of care, concern, and support from the people in east Texas was overwhelming and most appreciated.

That first day we received reports of large holes in the ground at Fort Polk, so I arranged for a helicopter and flew to the site. They were impressive holes. Whatever made them had fallen straight down at a tremendous speed. Contact with the Earth had splattered dirt twelve feet high on surrounding trees. The holes were three feet in diameter and very deep. Water was seeping into them, and the water level in the holes was seven feet below ground level. I could see enough of what was sticking above the water to know I was looking at parts of the pumps and power heads from the Shuttle's three main engines.

I showed up in Lufkin with only the few clothes I had with me at the Cape. We worked extremely long hours, and I visited laundromats at midnight to wash my clothes. It was the eighth or ninth day before I could dash home, see Karen, Amy, and Scott and his family, and get some more clothes. For the next three months I spent almost all of my time in Lufkin. At first we worked from the Lufkin Civic Center, but we soon moved to an empty floor in a bank building. The initial operation was focused on recovery of the crew, and I was not involved with that. Eight other astronauts helped to search for the crew and escorted their remains to the morgue. The recovery teams and the community at large conducted these solemn duties with the greatest sensitivity and reverence.

The debris recovery task was so enormous that it took us several days just to get our arms around the situation. Fellow astronaut John Grunsfeld and I calculated an estimated debris line, and we used that line to focus our search. Then astronaut Dom Gorie and I developed the search plan. We planned to conduct a ground search two miles to each side of the estimated debris line. Additionally, we would conduct a search from the air for three miles on each side of the ground-search area.

I was informed from the beginning that we wouldn't have the Texas Department of Public Safety and National Guard for very long. They pulled out when the crewmembers had all been found. Thankfully, Texas Forest Service Chief of Fire Operations Mark Stanford stepped in. Mark said he thought he could get a thousand firefighters to Texas in two days. I said, "Let's do it!"

Two days later we had one thousand firefighters initiating the systematic search for *Columbia* debris and more coming. Eventually, we had up to four thousand firefighters searching the fields and forests of east Texas at one time. "Boo" Walker from the Texas Forest Service coordinated the airborne search with about thirty-five helicopters and planes. Additionally, there was a fleet of about forty boats coordinated by US Navy Supervisor of Salvage and Diving Captain Jim Wilkins, looking through the numerous lakes and reservoirs along the debris line.

The earliest priority after finding the crew was anything from the leading edge of the left wing where the suspected damage from the foam strike occurred and everything that had recorded information on it. This included the Shuttle computers, which had a battery life of about thirty days before they would lose the contents of their memories. We also were looking for the orbiter experiments recorder that would give us information about all of the systems. We found the recorder on March 19 in almost perfect condition. It provided very useful information that had been recorded right up to the instant *Columbia* broke apart.

We looked for film from cameras that were mounted in the external tank disconnect area. These cameras could have photographed any foam that had broken free. We never found them.

We did recover some cassettes with videos the crew had taken of themselves on orbit. One of them showed the crew during an earlier part of their reentry. It was posted on the Internet and has been viewed by millions of people. I was glad we found that video. It showed the crew living large and having fun. There has been speculation NASA found video of the crew taken at the moment the Shuttle came apart, but that is not true. NASA shared everything we found.

The crew's personal items were the hardest to see. Any personal items or crew-worn items came to me for special processing. Searchers found the helmets the crewmembers had been wearing and the checklists they kept velcroed on their thighs. We found notebooks and many other items the crew had worn and carried. The items might come to us twisted, charred, and muddy, but I knew what they were.

We did get some interesting calls from the public. Someone found a bowling ball with "Columbia" printed on it. We assured the kind caller we hadn't launched any bowling balls into space.

Someone else found a little yellow dinosaur with purple spots on it. I looked through the entire flight manifest, but a stuffed dinosaur was not listed among the personal items the crewmembers took with them. I told the person who found it that it wasn't ours, but they sent it to us anyway. It was used as a mascot by the people working at Kennedy Space Center laying out *Columbia*'s parts on the hangar floor. And eventually, the little dinosaur did fly in space, twice, as a mascot for the STS-114 and STS-130 crews. STS-114 was the first Shuttle flight after the *Columbia* accident, and it meant a lot to all of us to see our little friend on board with Eileen Collins and her crew. The dinosaur now stands a silent vigil over the *Columbia* debris in the KSC Vehicle Assembly Building.

During the search, we continually asked ourselves questions. Had we calculated the right line for our search? Should we broaden our search? We were constantly analyzing what was being found and where it was being found to make sure we were directing the searchers to the right areas. The search was a huge undertaking. And it wasn't free.

FEMA was the federal agency funding all of the recovery efforts. One day FEMA Incident Commander Scott Wells came up to me and said, "We're spending almost $5 million a day on this." Then he put his finger into my chest and said, "You're responsible for 95 percent of that. When will we be done?"

I said, "Scott, we have all agreed to the search plan, and we are executing it as efficiently as the weather and the terrain will permit. We will be done when we are done."

I also worried about the cost. I grew up watching pennies, and I was concerned about the $5 million a day being spent. I'm a taxpayer, too, but we had a critical job to complete.

The firefighters came as fully-equipped, twenty-man teams and rotated in and out for two-week stays. The teams came from all around the country, and many firefighters were Native Americans. They worked with us for the full three months searching through fields, thickets, and swamps until the job was done. The firefighters, and in particular the Native Americans, were very proud to be helping our country's space program. In spite of the tragedy, it was a great feeling to have people from all over the country coming to aid the recovery. We wouldn't have succeeded without them.

I loved going out to the field camps at night, visiting with the searchers, signing autographs, and posing for photos. I assured them, "What you are doing is essential for NASA to get back into space. When we launch again, you guys will have played a big part in making that happen."

There is a small county seat in east Texas named Hemphill, and the folks there are incredible. They cooked thousands of meals for us for days while we searched for the crewmembers. They helped with those searches, and they kept working until we were done collecting debris. Sabine County and Hemphill continue to honor the *Columbia* crew with an annual memorial service, monuments, and a permanent exhibit, "Remembering *Columbia*," in the new Patricia Huffman Smith Museum. They are wonderful folks!

The debris from the Shuttle *Challenger* is now stored in two underground missile silos on Cape Canaveral Air Force Station. The *Columbia* parts are in a room in the Vehicle Assembly Building at the Kennedy Space Center.

Columbia launched in January 2003. December 2003 marked the one hundredth anniversary of the Wright brothers' first flights at Kitty Hawk, North Carolina. We had come so far in one hundred years, but we still have so much more to learn. A Shuttle didn't fly again until July 26, 2005—two and a half years after the *Columbia* accident.

After my seventh flight and the *Columbia* debris recovery, I went through a blue funk. Like an old baseball player walking to the dugout after a lifetime in the game, I realized it was over for me, and I would never fly again. I knew I'd had my last at bat.

I was disappointed and frustrated. I'm goal-oriented and always looking forward to what I'm going to do next. All I had ever looked forward to and all I had ever wanted to do was fly in space.

And now it was over.

But I stayed with the Space Shuttle program. I continued working as Chief of the Vehicle Integration Test Office, leading a group of about twenty engineers stationed at Kennedy Space Center and at Johnson

Space Center, who provided technical support to the Shuttle and ISS crews. The flight crews were busy training, so we routinely represented them at the hardware manufacturing facilities. The engineers were led by my right-hand man Jaime Forero and traveled to Japan, Europe, South America, Russia, and all over the US. They made sure all the hardware was safe and properly labeled, that interfacing pieces fit together, that NASA tools would fit on the equipment, and that information in the crew procedures was correct. They also supported Shuttle launch and landing preparations at Johnson, Kennedy, and Edwards.

I was as close to flying as I could be without leaving the ground. Among my responsibilities was being at the Cape for all practice countdowns, all launches, and all landings. As for *Columbia*, I was also responsible for the Kennedy Space Center Astronaut Crew Quarters. Now I was one of the NASA managers riding in the Astrovan part of the way to the pad with the crews on launch day.

When a Shuttle landed, I escorted the flight surgeons to the hatch as soon as it was opened so they could enter the orbiter and ascertain the physical condition of the crew. I stayed at the hatch and assisted each of the crewmembers out and to their feet in the small, white room immediately outside the hatch. I also directed the removal of certain crew equipment and other items that I wanted to maintain under my control and quickly return to Houston.

I stayed at NASA and in this job because I wanted to do everything I could to make sure my friends flew safely. I knew the managers of the Shuttle and ISS programs, and I had no problem talking to anyone at NASA if I thought there was something wrong. If during the countdowns I saw or sensed something I didn't like, I didn't have any qualms about sending an e-mail or calling the appropriate managers about my concerns. What were they going to do if they didn't like what I had to say? Tell me I couldn't fly anymore?

The NASA Engineering and Safety Center (NESC) was established after the *Columbia* accident. Its mission is to perform independent testing, analysis, and assessments of NASA's high-risk projects to ensure safety and mission success. As a second job, I served as the NESC Chief Astronaut from 2004 until 2007. I really enjoyed working with this extremely talented and dedicated group of engineers and scientists. They have made and continue to make valuable contributions across NASA, in line with their mission statement. The NESC has been called upon to apply their expertise to a wide range of challenges outside of NASA, including the rescue of the Chilean miners trapped underground in 2010.

The last Shuttle flew in July of 2011, and as sad I was to see that, it was time for the program to end. The Shuttle was designed for low Earth orbit space missions, and the time had come for us to turn our human spaceflight program toward deeper space—the Moon and Mars. As long as the Shuttles continued to fly, NASA would not have enough funds to leave low Earth orbit, return to the Moon, and eventually go on to Mars.

The Shuttles were also growing old. On nearly every mission we found additional issues about which we were concerned. If the Shuttles had continued to fly, statistically it would have been just a matter of time until we lost another vehicle and another crew.

But ending the Shuttle program has one major drawback. Since the retirement of the Shuttle, the only way for Americans or anyone to get to the ISS or to return from there to Earth is on a Russian Soyuz capsule.

In 2004, the Bush administration established the Constellation program, which included the plan for the retirement of the Shuttle. Constellation would put America on a path to deep space. Components of the Constellation program would also provide the United States with means of transporting our own and international partner astronauts to and from the ISS. The Constellation program was directed to establish a base on the Moon and to use that base to explore the Moon, to learn to mine and use its resources, and to develop the hardware and capabilities required for what will be a very long-duration human mission to Mars.

The plan included two new launch vehicles—called Ares I and Ares V. Ares I would launch crews into space aboard the new Constellation program Orion space capsule. Ares V would be much larger than Ares I and would launch the Altair lunar lander. After launch the

Orion would rendezvous with the Altair, and the upper rocket stage of the Ares V would send them to the Moon.

The Orion capsule would also be used to transport astronauts to and from the ISS. It was a good plan. The problem was the White House and Congress did not appropriate enough money to adequately fund development. This, in turn, caused the Constellation program schedule to slip and the costs to rise.

When President Barack Obama came into office, his administration canceled Constellation. The announcement hit the NASA space centers and the people who worked for NASA contractors like a ton of bricks. It was devastating. The atmosphere was similar to those after the *Challenger* and *Columbia* accidents. It was one of disbelief and devastation mixed with anger.

The new plan was for NASA to encourage and to pay commercial space transportation companies to build vehicles to shuttle astronauts to and from the ISS. Additionally, NASA eventually would travel to deep space after the discovery of "game-changing technologies," but there was no definitive plan for where to go, when to go, or how to get there.

When the entire Constellation program was canceled, billions of dollars the United States had spent developing it were wasted. After being canceled as part of the Constellation program, development and construction of the Orion capsule was eventually refunded. Work on Orion has resumed with the idea that it may be used initially as a backup system for transporting crews to the ISS. Any plans for Orion beyond that are unclear. Additionally, NASA was later redirected to again start working on a large, Ares V-category launch vehicle with a different name, Space Launch System, which the United States may use for something, someday, maybe.

NASA has a commitment to the International Space Station partners, so the United States is paying Russia to transport our astronauts to the ISS. That leaves our manned spaceflight program totally dependent and politically vulnerable. I frankly don't believe this is good for the United States, the international community, or the ISS program.

I am very upset with the current direction, or more appropriately, the lack of direction of the US space program. As demonstrated by the success of the Apollo program, NASA performs best when it is challenged with a specific goal and a definitive schedule, and is given enough money to get the job done. The NASA budget is currently less than one-half of one percent of the federal budget.

Forces at work inside NASA are dismantling much of the human spaceflight capabilities we used to have. Many of NASA's most talented managers and engineers have left the agency, taking their knowledge and years of experience with them. These forces are trying to push commercial spaceflight as far and as fast as they can, at the expense of and in place of NASA's own programs. Fortunately, some support in Congress is keeping NASA's human spaceflight program on life support until NASA management changes.

However, with the US budget and deficit pressures that now exist, the short-term prognosis for all of NASA's programs looks very bleak. NASA has cancelled almost all of its planned unmanned projects after the Mars Atmosphere and Volatile EvolutioN (MAVEN) spacecraft, which is scheduled for launch in late 2013. These have been one of the remaining bright spots in NASA's space exploration efforts. With these program terminations, NASA and the United States are breaking agreements with our international partners, and those partners are now turning to Russia and possibly China to establish new partnerships. The United States' leadership role in space exploration has been erased. Instead of leading, we are now following and falling behind the pack.

Have our dreams become too small, or have they just disappeared?

When I retired from NASA in January 2012, the administrator was Charles Bolden, a member of my ASCAN class. Charlie is a good person, and he has had a brilliant career. But we don't agree on what is happening at NASA.

Humans are inquisitive by nature. We want to know what's over the next hill. NASA can send a robot to Mars (if they have the money), but it's not the same as sending a person with eyes and a brain who can interpret and tell us what he or she has seen.

There is talk now within NASA that maybe in the years ahead we might send an American to an asteroid. I see no benefit in sending people to an asteroid. It's not a good idea. In my opinion, it would be a waste of money and time.

Apollo astronauts Neil Armstrong, Jim Lovell, and Gene Cernan testified to Congress against the Obama administration proposal to cancel the Constellation program. All of us in the Astronaut Office were cheering for them when they spoke. They spoke for us.

In response to a question of what he believed the US spaceflight priorities ought to be, Neil Armstrong responded:

1. Maintain US leadership in space
2. Guarantee US access to space
3. Continue to explore the solar system

I couldn't agree more!

Human space exploration not only expands our knowledge, but also stimulates innovation and industries that enhance our daily lives and drive economic development on our home planet.

Human space exploration engages the American workforce in developing new technologies required not only for meeting the challenges of leaving Earth, but also for meeting the challenges of a changing world.

I am very troubled about the future of our nation's space program. America spent more than fifty years investing in leadership in space with great technical, scientific, and economic dividends to our nation. And now we are walking away from that investment.

When will our country wake up?

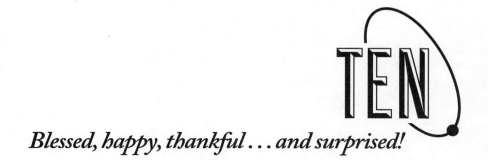

TEN

Blessed, happy, thankful ... and surprised!

I t's hard to put into words what a ship means to a sailor, what an aircraft means to a pilot, or what the Space Shuttle means to an astronaut.

They are much more than metal, wires, and switches. When you work for thirty years with a craft, you come to love it. You go through the same dangers and adventures together. You succeed and fail together. You work together to accomplish difficult assignments. Our Shuttles protected us from the unforgiving vacuum of space and brought us safely home. On the two occasions when they didn't, the vehicles had tried to warn us, but we failed to listen.

Our Shuttles—*Columbia, Challenger, Discovery, Atlantis,* and *Endeavour*—were more than spaceships. To those who prepared them for their voyages into space and for those of us who depended on them for our lives and knew that they had done everything we asked of them and more, they had souls. Each had its own unique personality. I flew five times on *Atlantis*, once on *Columbia*, and once on *Endeavour*. Guess which one is my favorite?

In the summer of 2011, the Shuttle program came to an end. It had fulfilled my childhood dreams. It had been my life for thirty-two years.

NASA invited astronauts, flight directors, and Shuttle program managers, former and active, to view the last Shuttle launch at the Kennedy Space Center, and each of us received four invitations. Karen, her friend Vicki Remsen, Amy, and her friend Karina Eversley excitedly accepted. The guests from Johnson Space Center stayed at a hotel in Orlando, and the night before the launch there was a big party at the hotel.

As usual, I was working at Astronaut Crew Quarters and was going to be very busy the next morning, so I didn't think I could attend. But at nearly the last minute I decided I didn't want to miss the party. Without telling Karen and Amy my plans, I arrived at the party before they did and surprised them when they came through the door. It was a nice evening. I saw people I hadn't seen in twenty-five years. There were hundreds of people in attendance, including about eighty-five astronauts. It was great to see everyone there from the decades of the Shuttle program. The evening brought back many wonderful memories.

The final Shuttle mission was on *Atlantis*. It was the 135th flight in the history of the program—STS-135. It was very difficult for all of us to say good-bye to the Shuttles, to the program, and in many cases to each other. People were already retiring and leaving NASA, and thousands were being laid off.

The morning of the launch, STS-135 crewmember Rex Walheim and I met for prayer, just as we had when we flew together on STS-110, and just as we and Leland Melvin had before Rex and Leland flew on STS-122.

In spite of challenging weather conditions and a brief dramatic hold at T minus 31 seconds to visually verify that the gaseous oxygen vent arm was fully retracted, *Atlantis* launched at 11:29:03.9 a.m. EDT on July 8, 2011.

After the Shuttle had disappeared from our view and safely reached orbit, NASA and its contractors had a big celebration in the Vehicle

Assembly Building—hot dogs and beer. Everyone was invited, and I attended. I wore my blue flight suit. I wasn't interested in eating; what I wanted to do was to thank the people whose hard work, attention to detail, and dedication made the Shuttles fly for so many years. There were long tables set up with food, and people stood in two lines waiting their turn. I started at the tables and walked back down the line. I introduced myself to every person, talked to them, and shook their hands. I posed for photos and thanked them for their years of service and all the great things they had each done. I talked to more than five hundred people that day, all of them extremely proud about what they, individually and collectively, had accomplished. It was a very emotional time for me and everyone associated with the program.

In Houston two nights before *Atlantis* was scheduled to land, I was so wired that I only got about three and a half hours of sleep. That morning Karen and I went to the airport. She flew to Indiana to visit her folks, and I flew to Orlando. I arrived at the Kennedy Space Center around three o'clock in the afternoon. At five o'clock I tried to lie down and get some rest because I had to be up a little before midnight to start preparations for landing, but I couldn't sleep. I got up at midnight and worked straight through until six o'clock that night before finally collapsing.

Minutes before six o'clock in the morning on July 21, 2011, I was standing on the Shuttle runway waiting to greet the crew of the final mission of the Space Shuttle program. It was the same place I had stood and waited for the arrival of the *Columbia* crew, and many others. A large contingent of NASA officials waited nearby to say hello to the returning voyagers and good-bye to the Shuttle program.

It was a beautiful, early morning in Florida, before first light, before the heat of the day arrived. Fresh, clean air blew in lightly off the ocean. We saw the International Space Station pass overhead, and we watched it travel across the sky for about four minutes. Then, right on cue, *Atlantis* announced its arrival with the Shuttle's signature double sonic boom. I strained to find *Atlantis* as she dropped quickly and silently toward the runway at the break of dawn. Just minutes later the orbiter came down into the runway lights.

All of us standing on the ground listened on radios as we witnessed the last landing of a Space Shuttle.

"The pre-flare maneuver executed," said Rob Navias, the STS-135 entry commentator in Mission Control, Houston. "Landing gear down and locked . . . main gear touchdown . . . nose gear touchdown . . ."

"Having fired the imagination of a generation," he continued, "a ship like no other, its place in history secured, the Space Shuttle pulls into port for the last time. Its voyage at an end."

The astronaut working as CAPCOM that morning, Barry "Butch" Wilmore, put an exclamation mark on the success of the Space Shuttle program: "Job well done, America!"

They were good words, emotional words. STS-135 Commander Chris Ferguson radioed his comments from his seat inside *Atlantis:* "The Space Shuttle has changed the way we view the world, and it's changed the way we view our universe. There's a lot of emotion today, but one thing is indisputable: America is not going to stop exploring. Thank you *Columbia, Challenger, Discovery, Endeavour,* and our ship, *Atlantis,*" he said. "Thank you for protecting us and bringing this program to such a fitting end. God bless all of you. God bless the United States of America."

It was a beautiful landing and a perfect ending. It was a great feeling to see *Atlantis* come to a stop on the runway and to welcome the crew home, safe and sound. After helping the crew out of *Atlantis,* I crawled into the orbiter for the last time, collecting items and making sure nothing was left behind. My job had enabled me to be in *Discovery, Endeavour,* and *Atlantis* on the launch pad prior to their final flights and to be in them again after they landed. I felt very much at home on board these ships. But now I had an empty, lonely feeling inside me. It was like losing a good friend.

What did the Space Shuttle program accomplish? The Space Shuttle made it possible for our country to utilize low Earth orbit for scientific and engineering research, astronomy, environmental studies, and communications. It allowed us to repair satellites and to build the International Space Station. It opened space to nonpilots, women, and minorities. It brought nations together, working for peaceful purposes.

On April 12, 2011, the thirtieth anniversary of the first Shuttle flight, NASA Administrator Charles Bolden announced the facilities where NASA's four orbiters would be permanently displayed for the public.

Enterprise, the first orbiter built and never intended to fly in space, moved from the Smithsonian's National Air and Space Museum Steven F. Udvar-Hazy Center in Virginia to the Intrepid Sea, Air and Space Museum in New York City. *Discovery* was placed at the Steven F. Udvar-Hazy Center near Washington Dulles International Airport. *Endeavour* went to the California Science Center in Los Angeles. *Atlantis* remained in Florida at the Kennedy Space Center Visitor Complex.

The Shuttles each had a place to go, but most of the Shuttle program workers did not. Thousands of people on the Florida Space Coast, in Houston, Louisiana, Utah, Alabama, and elsewhere around the United States were laid off from their jobs because the Shuttle program had come to an end and there was nothing ready to replace it. These were people who just like me had dedicated their lives to NASA, the Shuttles, and space exploration. They loved the Shuttles as much as I did. I miss my fellow space program workers and the excitement and challenges of what we did together.

I was fortunate to be able to live a relatively normal life while having the opportunity to fly in space. The Shuttle program was never like Mercury, Gemini, and Apollo, when TV trucks lined up in front of astronauts' houses and the media filled their lawns. We didn't live under the media microscope, but we did make a lot of public appearances.

The Astronaut Office receives more than two thousand requests each year for astronauts to make appearances, but less than 50 percent of them can be accommodated. While it was not fun to pack and leave home and family behind again and again, in aggregate I enjoyed making the appearances. I have enjoyed seeing new places, meeting people, and informing them about our country's space program.

I also enjoyed opportunities to meet celebrities. Over the years I met Charles Barkley, Greg "Pappy" Boyington of the "Black Sheep Squadron," Drew Brees, James Brolin, Jimmy Buffett, Roy Campanella, David Crosby, Scott Crossfield, Tom Cruise, Penélope Cruz, His

Holiness the Fourteenth Dalai Lama, Garfield creator Jim Davis, John Denver, Hugh Downs, Joe Frazier, James Garner, Clarence Gilyard, Jr., David Hasselhoff, Florence Henderson, the Judds, Art Linkletter, Kenny Loggins, Don Mattingly, Scotty McCreery, Roger E. Moseley, Jim Nabors, Graham Nash, Leonard Nimoy, Peter Noone, Sandi Patty, Roger Penske, *Charlie's Angels* actress Jaclyn Smith, Tom Smothers, Stephen Stills, Joe Torre, and Al Unser.

There are times when I've been treated like a celebrity; however, sometimes the celebrity treatment can be somewhat dubious. Every year the College of Agriculture at Purdue University holds the Ag Alumni Fish Fry. One year I was invited as an honored guest. Before the program, Maury Williamson, the executive secretary of the Agricultural Alumni Association, had an outhouse put on stage.

In front of several thousand people, Maury had me walk into the outhouse and close the door. As the alumni looked on, he detonated some explosives and "launched" the outhouse to the Armory ceiling. The audience roared with laughter. Fortunately, Maury had given me a trapdoor, and I escaped before the "liftoff."

In 1997, I was invited to a science fiction writer meeting in Beijing, China. Karen wasn't able to go, so I took Mom with me. Fellow astronaut Shannon Lucid, who was born in Shanghai to American Baptist missionary parents, was on the trip too, and we joined three Russian cosmonauts, including my friend General Alexei Leonov. The stage was set for a really good time.

Taking my mother to China turned out to be a very smart thing to do. The Chinese respect and honor their elders, and they thought bringing my mother was a sign of my good character. I was being a very good son!

Mom had a wonderful time. She got to see the Great Wall, the Forbidden City, giant pandas, and lots more. The only problem with the trip was that Shannon and Mom instantly struck up a strong friendship. Shannon became my surrogate sister, and she and Mom teamed up against me. They enjoyed pestering me, and I had to give it back to them. Or maybe I pestered them first, and they gave it back to me. It's hard to remember details.

As we traveled around China, we were greeted graciously by city leaders and company officials wherever we went. One visit is especially memorable. We stopped at a company that made drinks out of algae. We tried the beverage, but it was definitely an acquired taste. After the plant tour, our hosts provided an elaborate, traditional Chinese meal for us, and I had a difficult time finding something that was appealing to me. I did find a bowl of broth that I really liked. I was enjoying the broth when my interpreter noticed that I liked it. She told me that there was a special treat for me in the broth. I had no idea what she meant, and I must have looked puzzled.

"No, you don't understand," the interpreter said. She stuck her chopsticks into the bowl of broth and pulled out a little frog. A whole cooked frog. It had teeth!

"These are very good, and they have honored you by putting it into your bowl," she explained.

"I am honored," I said as graciously as I could. "Please tell them, 'Thank you very much.' Would you like to have it?"

Her eyes lit up. "Oh, thank you!" she said. She popped the frog into her mouth, and it crunched as she ate it.

Mom loved the trip; she didn't get a frog.

Mom was diagnosed with cancer at Christmastime in 2001. She was not feeling well, but she came to the Cape for my last flight in April 2002. The day before each launch, NASA brought buses of family members and guests to an area near the launch pad. The crew arrived, stood at a distance from them, and waved and talked. While we were there, Mom fainted. I saw her go down, but there was nothing I could do because of the quarantine restrictions. That was very hard. My sisters, Janet Rattazzi and Judi Futa, were right at Mom's side with other family members. My best friend, Dr. Jim Gentleman, was there too. They took good care of her, and she was fine, but I felt very helpless.

In late February 2009, Mom left the hospital and went into hospice care at home, where she wanted to be. Judi took time off from her teaching job to be with her full time. Mom hoped that she would live until June when she would turn eighty-two, because both her parents had lived to that age. When I went back home to see her, my sisters were treating every day like Mom's birthday. I said, if we were doing that, we should have a birthday party. We invited family and friends, I cooked one of Mom's favorite meals, and we made a birthday cake. The birthday party was a nice evening, and Mom had a wonderful time surrounded by her family and friends. Janet and I spent much of Mom's last month with her. All three of us were there around her bed when she passed away on March 24, 2009. We were happy we had celebrated her birthday early.

My sisters and I sold the house that Dad had built and that Mom lived in for more than sixty years. It was very hard to lose Mom and to sell our family home. Everything was changing.

I had decided to stay at NASA through the end of the Shuttle program and then retire. I made that decision well before all of the turmoil that came with the termination of the Constellation program and NASA Headquarters' push for commercial manned spaceflights began. As the Shuttle program was winding down, Karen was laid off and retired from her position with the Shuttle contractor United Space Alliance. Nearly sixteen months later, I retired from NASA on January 20, 2012, my sixty-fourth birthday.

Retiring when I did was the right time for me, but as my final day at NASA approached, I had some feelings that I was a rat leaving a sinking ship. I care very deeply about our country and firmly believe that a strong national space program strengthens the United States. Much of the technology that powers our economy today was developed by or for the space program. Spending money on advanced space technologies is like growing seed corn for our future economy. Looking back at history, when civilizations, empires, and countries stopped exploring and innovating, that is when they started to decline. I am afraid for our country that this is the juncture where we currently find ourselves.

Now that a little time has passed, I find a sense of relief in retirement for two reasons.

First, the Shuttle program was always the elephant in the room. Any time we wanted to make family plans, I had to first check the Shuttle schedule. For thirty years, whatever we wanted to do had to fit around those dates. Additionally, we had to plan what we were going to do if the schedule changed, and the schedule was always changing. We ended up cancelling or adjusting a lot of family plans. That pressure is gone now, and it feels liberating.

Second and most importantly, there was a nagging feeling in the pit of my stomach, especially after the *Columbia* accident. I was concerned for the safety of my friends who were flying. NASA was doing everything it could to make sure the vehicles were as safe as they could be, constantly asking, "What are we missing? What are we not smart enough to see? Is there something the vehicle is trying to tell us—again?" But if we kept flying the Shuttles, it was most likely only a matter of time before another tragedy would have occurred. I'm happy to no longer have that feeling in my gut.

Karen and I are looking forward to an active and fun retirement. We're heading out to see America and the world. I've seen it from space; now we want to see it up close and personal. We want to spend time with our family and enjoy watching our granddaughters, Cassidy, Katie, and Emily, grow up. Karen and I are looking forward to participating in more, and maybe eventually leading, mission trips to share God's message of grace.

I am looking forward to renewing my efforts on our families' genealogies and updating the four genealogy books that I have already published. Pursuing our ancestry will lead us all over the US and possibly back to Europe.

I'd like to see a ball game in every Major League Baseball park in the country, and I've got a pretty good start toward that goal. Karen and I have talked about going to different parts of the country, renting an apartment for a month or two, and doing day trips from that home base. I want to stay in some small town in New England during the fall and relax among the beautiful changing colors.

I still do speaking engagements. I especially enjoy getting young people excited about space—just like I was at their age and still am today. I like talking to students and encouraging them to use their God-given talents, to set goals, to study hard, to work hard, and to not give up too easily. I tell them their dreams can come true.

We need to continue to excite young people. We need to challenge them with specific national space goals with demanding but achievable schedules, and support them with our national resolve and funding.

In 2003, the Crown Point Community School Corporation built a new elementary school, and they asked the parents what they wanted it to be named. The parents chose "Jerry Ross Elementary School." What an incredible honor! The school has third, fourth, and fifth grade students, and I visit to tell the students my story and encourage them to follow their dreams, just like I did.

———————————————●———————————————

SCOTT ROSS: *Through my dad and then my mother, my whole life has been wrapped up in NASA and the Space Shuttle. I was too young to remember anything about Wright-Patterson Air Force Base in Ohio. I remember a little about Edwards—going to the desert and the lizards and snakes we saw. I remember the sonic booms; they were pretty regular. I remember when the Shuttle Enterprise landed and going with my dad to watch it. I know my dad took a lot of photos of the B-1 while we were there. Our family albums have one picture of us, four photos of the B-1, another shot of us, and five more photos of the B-1 . . .*

I remember being at my grandparents' home in Indiana when Dad called to tell us he'd been selected as an astronaut. We'd been out in the barn, and they called us into the house. I was excited.

We went to all of Dad's launches, and they were fun. It was like going to a family reunion because everyone was there. We'd stay in a condo and see the launch. Sometimes we went to Disney World. The people at the Kennedy Space Center treated the families really special. We got to see and do things most people never experienced.

I remember the first launch very well. I wasn't prepared for it. I was excited and scared, and I really didn't know what was going on. I had no idea how powerful it was going to be. It was overwhelming and just something you couldn't prepare yourself for. I cried. And to be honest, I teared up at every one of Dad's launches.

When I was a kid, my dad never seemed out of the ordinary to me. Being an astronaut was his job. It was pretty special to have a dad doing all this stuff, but still, he was plain old Dad. He'd come home from a spaceflight, and he'd be out mowing the yard the next day. You don't think about astronauts coming home from space and mowing grass.

I asked him all the time what it was like. We'd sit down and talk, and I'd listen to him explain everything. He always said it was an amazing experience, something you really can't fully understand unless you've been there.

The only thing I've really ever known is having an astronaut for a dad. He's a great dad, and now he's also a wonderful grandfather. It's been very special to grow up like this. NASA and the Shuttle program have been a big part of our lives—like another member of the family. Now I'm sad to see it go. I'm sad for my dad, because he's come to a natural stopping point in his career. He says he was ready to retire, but I know how much all of this meant to him. It's very hard.

The older I get, the more I realize how different it was to grow up with a dad who flies in space and goes to the White House. How many kids can say that?

I think about what I'm going to tell my grandchildren someday when they ask about my dad and what he did. There's a whole lot to say. It's hard to put everything he accomplished and everything I feel into words.

I guess I'll say, to me, he's a real American hero.

President Reagan was right when he said in his 1984 State of the Union Address, "America has always been greatest when we dared to be great. We can reach for greatness again. We can follow our dreams to distant stars, living and working in space for peaceful, economic, and scientific gain." Those words are still true today and should be used as our guide to the future.

I believe humans will someday walk on Mars. It will probably not be in my lifetime, which surprises me. Early in my astronaut career, I actually thought I might have a decent chance of going to Mars. We will get there someday because no one can stop the human need to explore.

I feel blessed to have served our country in such a unique and exciting way. I guess that one word—blessed—is the best way to sum up how I feel about the entire experience. I'm also happy I was able to follow my dream. I'm pleased and so thankful. And now looking back on it all, I am really surprised at everything God allowed me to do.

The Shuttle program is done, but sometimes at night, I look up at the starry sky. I see the Moon, stars, and sometimes I see the International Space Station—a white dot of light moving rapidly across the blackness above me. When I see that dot, it makes me smile to know that I am watching a place where I've been, a place I helped build, a place where my friends are conducting scientific research every day, and a place where I wish I was going again.

When I look up at the sky on those nights, I think about flying in space. I think about Jim Gentleman and me lying on hay bales in Grandpa Joe's field next to my house in Crown Point. I remember how cool I thought it would be to go up there and be among the stars.

Today I'm on the other side of the dream. I'm looking back on it and realize that I was absolutely right about it all.

Except for one thing.

Working in our country's space program and flying in space were even better and even more rewarding than I ever dreamed.

MY JOURNEY

A timeline

- Born January 20, 1948.
- October 4, 1957, *Sputnik 1* was launched by the Soviet Union, and *Explorer 1*, the first US satellite, was launched on January 31, 1958. I was in the fourth grade and decided to go to Purdue University, to become an engineer, and to get involved in our country's space program.
- April 12, 1961, Soviet Cosmonaut Yuri Gagarin became the first human to fly in space.
- May 5, 1961, Alan Shepard became the first American to fly in space.
- May 25, 1961, President John F. Kennedy announced the goal of landing a man on the Moon in a speech to a Joint Session of Congress.
- June 16, 1963, Soviet Cosmonaut Valentina Tereshkova became the first woman to fly in space.

- March 18, 1965, Soviet Cosmonaut Alexei Leonov conducted the first spacewalk.
- September 1966, I entered Purdue University.
- July 20, 1969, Neil Armstrong and Edwin "Buzz" Aldrin, Jr., landed on the Moon.
- January 25, 1970, I married Karen Sue Pearson.
- June 1970, commissioned a second lieutenant in the United States Air Force and graduated with a bachelor of science degree in mechanical engineering.
- March 30, 1971, our daughter, Amy Jo, was born in Home Hospital, Lafayette, Indiana.
- January 1972, received my master of science degree in mechanical engineering.
- February 15, 1972, I entered active duty in the US Air Force at Wright-Patterson Air Force Base, Ohio.
- April 27, 1972, our son, Scott, was born at Wright-Patterson Air Force Base, Ohio.
- August 1975, reported to Edwards Air Force Base, California, to attend the Flight Test Engineer course at the US Air Force Test Pilot School.
- July 16, 1976, I graduated from the US Air Force Test Pilot School. Apollo 11 Command Module Pilot Michael Collins was the guest speaker.
- August 1976 through February 1979, served as a Flight Test Engineer on the B-1 bomber Joint Flight Test Team.
- February 1979, transferred to NASA Johnson Space Center, Houston, Texas, to work as a US Air Force detailee helping to integrate military payloads into the Space Shuttle.
- May 1980, selected as a NASA astronaut.
- April 12, 1981, the first Space Shuttle flight, STS-1, launched.
- November 26 through December 3, 1985, my first flight, STS-61B, on Space Shuttle *Atlantis*. The crew included Commander Brewster Shaw, Pilot Bryan O'Connor, Mission Specialists Mary Cleave and Sherwood Spring, and Payload Specialists Charles Walker and Rodolfo Neri-Vela. The crew launched three communications

satellites, and I performed two spacewalks to evaluate manually building structures in space.

- December 2 through December 6, 1988, my second flight, STS-27, a classified military mission, on *Atlantis*. The crew included Commander Robert "Hoot" Gibson, Pilot Guy Gardner, and Mission Specialists Mike Mullane and Bill Shepherd.
- April 5 through April 11, 1991, my third flight, STS-37, on *Atlantis*. We deployed the Compton Gamma Ray Observatory (CGRO), and I performed an unplanned spacewalk to repair the $670 million CGRO as well as a planned spacewalk to evaluate hardware for the International Space Station. The crew included Commander Steve Nagel, Pilot Ken Cameron, and Mission Specialists Linda Godwin and Jay Apt.
- April 26 through May 6, 1993, my fourth flight, STS-55, on *Columbia*, a German-sponsored Spacelab mission during which we conducted eighty-eight experiments. The crew included Commander Steve Nagel, Pilot Tom Henricks, Mission Specialists Charlie Precourt and Bernard Harris, Jr., and German Payload Specialists Hans Schlegel and Ulrich Walter.
- November 12 through November 20, 1995, my fifth flight, STS-74, on *Atlantis*. We added a docking module to the Russian Mir space station. The crew included Commander Ken Cameron, Pilot Jim Halsell, and Mission Specialists Bill McArthur and Chris Hadfield.
- December 4 through December 15, 1998, my sixth flight, STS-88, on *Endeavour*. We started the assembly of the International Space Station by rendezvousing with the orbiting Russian *Zarya* module and joining it to the US *Unity* module. I performed three spacewalks to assemble the modules and to mount equipment on their exteriors. The crew included Commander Bob Cabana, Pilot Rick "CJ" Sturckow, and Mission Specialists Jim Newman, Nancy Currie, and Sergei Krikalev.
- April 8 through April 19, 2002, my seventh and last flight, STS-110, on *Atlantis*. We added the S-zero truss element to the International Space Station. I performed two of the four spacewalks to attach S-zero. The crew included Commander Mike Bloomfield,

Pilot Steve Frick, and Mission Specialists Ellen Ochoa, Steve Smith, Rex Walheim, and Lee Morin.

- 2003 through 2011, I served as the Chief of the Vehicle Integration Test Office, leading a group of engineers who provided technical support to Space Shuttle and International Space Station crews by representing them at hardware manufacturing facilities around the world and otherwise assisting in their preparations for flight.

- February through April 2003, helped lead the *Columbia* debris recovery efforts in east Texas.

- 2004, presented the Distinguished Engineering Alumnus Award by Purdue University's College of Engineering.

- July 21, 2010, the last Space Shuttle flight, STS-135, landed at the Kennedy Space Center, Florida.

- January 20, 2012, I retired from NASA.

STS-61B

Anatomy of a launch

STS-61B was the second trip to space for Shuttle Atlantis. *It was launched from Kennedy Space Center in Florida at 7:29 p.m. EST on November 26, 1985. The Shuttle crew deployed three communications satellites and tested construction techniques in orbit.* Atlantis *landed at Edwards Air Force Base, California, at 4:33 p.m. EST on December 3, 1985, after six days and twenty-one hours in orbit.*

Launch-1 week. Astronauts go into quarantine to prevent the contraction of colds or illnesses.

Launch-3 days. Astronauts arrive at Kennedy Space Center. The Shuttle's payload bay has already been loaded and closed. The spouses of the astronauts are escorted to the quarantine facility for a traditional steak dinner and brief visit.

Launch-2 days. NASA hosts a barbecue for the crew, who each invite four or five adult guests. The guests must be checked by a flight surgeon before they arrive at the Beach House.

Launch-1 day. The Astronaut Support Personnel, or "Cape Crusaders," set up the cockpit of the Shuttle. They install the cue cards, checklists, and all of the equipment needed for launch. This includes seats, communications lines, and emergency air supplies. They set every one of more than 1,000 switches and circuit breakers in the correct position for launch and double- and triple-check them. The astronauts' families host prelaunch receptions in venues all over Cape Canaveral.

Launch-8 hours. The launch pad is cleared of all personnel in order to safely fill the large, orange external fuel tank. Five hundred thousand gallons of liquid oxygen and liquid hydrogen will be needed for lift off.

The "T-" times are the countdown clock times. Since the countdown includes planned pauses, it doesn't indicate the actual time left until launch until the countdown clock is started at nine minutes before launch. At that time there are no more planned pauses in the countdown; therefore, without any unplanned halts to the countdown, the launch time will be as indicated by the countdown clock.

T-4:20:00. Astronauts are awakened in the crew quarters. The Shuttle is now fully fueled. They have breakfast in front of the cameras, dress in their flight suits, and attend a briefing on the Shuttle's status and the weather.

T-2:50:00. The astronauts wave to crowds of well-wishers as they climb into the Astrovan for the twenty-minute ride to the launch pad. Accompanying them are a driver, security personnel, the Chief of the Vehicle Integration Test Team, and an insertion technician. Other NASA personnel may join them; for STS-61B, John Young and George Abbey rode as far as the airstrip from which NASA weather aircraft would take off and the Launch Control Center, respectively.

T-2:30:00. The astronauts arrive at the launch pad. They ride an elevator 195 feet up, to the hatch into the Shuttle. A seven-person closeout crew helps them strap in, closes and seals the hatch, and then moves to a safe distance at least three and a half miles away. By this time, the spouses and children of the crew are in the Launch Director's office at the Launch Control Center, marking time.

T-0:09:00. The countdown enters a twenty-minute hold. This is the last chance for the NASA launch team to make sure the restricted area around the launch pad is clear and that everything is ready for launch. After the twenty-minute hold, the countdown resumes. The crew's families are directed to the roof of the Launch Control Center, from which they can watch the launch. Additional guests are in bleachers outside the Launch Control Center.

T-0:05:00. The pilot starts the Auxiliary Power Units.

T-0:00:31. Control of the countdown is transferred from the Launch Control Center to the Shuttle's onboard computers.

T-0:00:07. A valve on a nearby water tank opens, flooding the mobile launch platform and the flame trench at the base of the pad with 300,000 gallons of water. This water will protect the pad and Shuttle from heat and acoustic damage.

T-0:00:06.6. The orbiter's three main engines ignite, drawing fuel from the external tank. The astronauts can feel the vibrations caused by the engines.

T-0:00:00. The solid rocket boosters on the sides of the external tank ignite. Simultaneously, eight nuts on the huge bolts holding the Shuttle to the launch pad are fractured using explosives. The astronauts can feel the sudden acceleration like a blow to the back of their seats. The Shuttle has now lifted off the launch pad.

T+0:00:40. The Shuttle has accelerated quickly. It is already going faster than the speed of sound.

T+0:01:00. The Shuttle goes through maximum dynamic pressure, the point where the speed of the orbiter and the density of the atmosphere combine to create the largest aerodynamic forces on the Shuttle stack. The engines briefly throttle down to keep the pressure from damaging the vehicle.

T+0:02:00. The two solid rocket boosters have used up all their fuel. They are jettisoned. The Shuttle is already twenty-five miles high and traveling at four times the speed of sound.

T+0:07:30. The Shuttle begins to level off its flight and accelerates at three g's. This is uncomfortable for the astronauts, who feel a lot of pressure against their bodies.

T+0:08:30. The Shuttle's main engines shut down. The external tank is jettisoned; it will be destroyed as it reenters Earth's atmosphere over the Indian Ocean. The Shuttle is about one hundred miles high and traveling at almost 17,500 miles per hour.

T+0:48:30 (Approximate). A final rocket firing from the smaller Orbital Maneuvering System engines place the Shuttle into its final circular orbit. The Shuttle will circumnavigate the globe every ninety minutes.

Learn more about the life and work of Jerry Ross as well as other Purdue University astronaut alumni and flight pioneers . . .

The Barron Hilton Flight and Space Exploration Archives reflects Purdue University's strong ties to the history of aviation and space exploration. Through digitization of original collections, creation of physical and virtual exhibitions, lectures and presentations, the Archives serves as a critical, dynamic educational resource on flight and space history for scholars, students, teachers, and the public.

Established with generous funding from the Conrad N. Hilton Foundation and Barron Hilton, the Archives include the papers, photographs, and rare books of astronauts, pilots, engineers, and promoters of flight and space travel. Some of the key collections in the Archives include papers of astronauts Neil Armstrong, Roy Bridges, Jr., Eugene Cernan, Jerry Ross, and Janice Voss; papers of pilots such as Amelia Earhart, first woman to fly solo across the Atlantic Ocean, and Ralph Johnson, inventor of the technique for safe landings during blackout conditions, and papers of engineers whose contributions advanced space exploration, such as Mark Geyer, Orion program manager for NASA.

Purdue astronauts gather for a 1999 reunion at the University.

To view objects, documents, and see videos and interviews, visit: www.lib.purdue.edu/spcol

If you enjoyed Spacewalker, *the book, you may be interested in the enhanced digital versions . . .*

Available for most e-book readers: *Spacewalker, the Enhanced e-Book* (ISBN 978-1-61249-233-9)

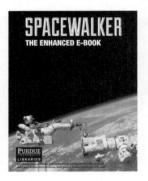

- The complete text of *Spacewalker,* the book

- Almost thirty videos, documenting Jerry Ross's career at NASA with commentary

- Over fifty images, also with commentary by Jerry Ross

Available for iPad in the Apple App Store: *Spacewalker, the App*

- A dynamic timeline of Jerry Ross's life and the Space Shuttle's history

- An interactive quiz

- Almost thirty videos, documenting Jerry Ross's career at NASA with commentary

- Over fifty images, also with commentary by Jerry Ross

- The complete text of *Spacewalker,* the book

You may also be interested in visiting the authors' websites at www.jerrylross.com and www.johnnorberg.com.

Further information about other books on space and aviation history, and other publications from Purdue University Press, can be found at: www.thepress.purdue.edu